LAB AND SAFETY SKILLS
IN THE SCIENCE CLASSROOM

GLENCOE
McGraw-Hill

New York, New York   Columbus, Ohio   Woodland Hills, California   Peoria, Illinois

Copyright © The Glencoe/McGraw-Hill Companies, Inc. All rights reserved. Permission is granted to reproduce the material contained herein on the condition that such material be reproduced only for classroom use, be provided to students without charge, and be used solely in conjunction with Glencoe Science programs. Any other reproduction, for use or sale, is expressly prohibited.

Send all inquiries to:
Glencoe/McGraw-Hill
936 Eastwind Drive
Westerville, OH 43081

ISBN 0-02-826853-9

Printed in the United States of America

11 12 13 14 15 16   066   05 04 03 02 01 00 99

TABLE OF CONTENTS

Introduction 5
International System of Units 5
Laboratory Safety 7
Safety Symbols 9
Safety Rules 10
Safety Contract 10
Laboratory Design and Equipment 11
Evaluating Activity Work 12
Laboratory Techniques 14
Laboratory Skills Assessment 22
    Scientific Problem Solving 22
        Scientific Methods 22
        Tables and Graphs 23
    Safety in the Science Classroom 26
    Using a Microscope 30
    Measuring with a Microscope 31
    Using Laboratory Measuring Devices 32
    Using a Metric Ruler and Protractor 34
    Estimating Values Using SI 36

INTRODUCTION

Science is the body of information including all the hypotheses and experiments that tell us about our environment. All people involved in scientific work use similar methods of gaining information. One important scientific skill is the ability to obtain data directly from the environment. Observations must be based on what actually happens in the environment. Equally important is the ability to organize this data into a form from which valid conclusions can be drawn. The conclusions must be such that other scientists can achieve the same results.

INTERNATIONAL SYSTEM OF UNITS

The International System (SI) of Measurement is accepted as the standard for measurement throughout most of the world. Three base units in SI are the meter, kilogram, and second. Frequently used SI units are listed below.

TABLE 1

	Frequently used SI units
Length	1 millimeter (mm) = 1000 micrometers (μm) 1 centimeter (cm) = 10 millimeters (mm) 1 meter (m) = 100 centimeters (cm) 1 kilometer (km) = 1000 meters (m) 1 light-year = 9 460 000 000 000 kilometers (km)
Area	1 square meter (m²) = 10 000 square centimeters (cm²) 1 square kilometer (km²) = 1 000 000 square meters (m²)
Volume	1 milliliter (mL) = 1 cubic centimeters (cc) (cm³) 1 liter (L) = 1000 milliliters (mL)
Mass	1 gram (g) = 1000 milligrams (mg) 1 kilogram (kg) = 1000 grams (g) 1 metric ton = 1000 kilograms (kg)
Time	1 s = 1 second

Temperature measurement in SI are often made in degrees Celsius. Celsius temperature is a supplementary unit derived from the base unit kelvin. The Celsius scale (°C) has 100 equal graduations between the freezing temperature (0°C) and the boiling temperature of water (100°C). The following relationship exists between the Celsius and kelvin temperature scales:

$$K = °C + 273$$

Several other supplementary SI units are listed below.

TABLE 2

Supplementary SI units			
Measurement	**Unit**	**Symbol**	**Expressed in base units**
Energy	joule	J	kg · m²/s²
Force	newton	N	kg · m/s²
Power	watt	W	kg · m²/s³ or J/s
Pressure	pascal	Pa	kg/m · s² or N · m

TABLE 3

SI/Metric to English conversions			
	When you want to convert:	Multiply by:	To find:
Length	inches	2.54	centimeters
	centimeters	0.39	inches
	feet	0.30	meters
	meters	3.28	feet
	yards	0.91	meters
	meters	1.09	yards
	miles	1.61	kilometers
	kilometers	0.62	miles
Mass and weight*	ounces	28.35	grams
	grams	0.04	ounces
	pounds	0.45	kilograms
	kilograms	2.20	pounds
	tons	0.91	tonnes (metric tons)
	tonnes (metric tons)	1.10	tons
	pounds	4.45	newtons
	newtons	0.23	pounds
Volume	cubic inches	16.39	cubic centimeters
	cubic centimeters	0.06	cubic inches
	cubic feet	0.03	cubic meters
	cubic meters	35.31	cubic feet
	liters	1.06	quarts
	liters	0.26	gallons
	gallons	3.78	liters
Area	square inches	6.45	square centimeters
	square centimeters	0.16	square inches
	square feet	0.09	square meters
	square meters	10.76	square feet
	square miles	2.59	square kilometers
	square kilometers	0.39	square miles
	hectares	2.47	acres
	acres	0.40	hectares
Temperature	Fahrenheit	5/9 (°F − 32)	Celsius
	Celsius	9/5 °C + 32	Fahrenheit

*Weight as measured in standard Earth gravity

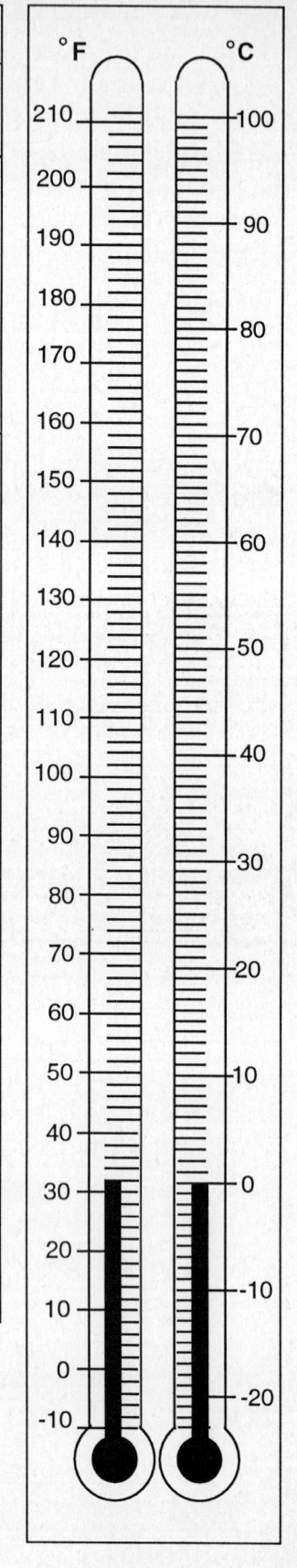

LABORATORY SAFETY

Outlined below are some considerations on laboratory safety that are intended primarily for teachers and administrators. Safety awareness must begin with the principal, be supervised by the department head, and be important to the individual teacher.

Principals and supervisors should be familiar with the guidelines for laboratory safety and provide continual supervision to ensure compliance with those guidelines. Principals and supervisors need to be supportive of teachers requesting assistance with implementing safety procedures. Teachers have the ultimate responsibility for enforcing safety standards in the laboratory. They must set the proper example in the laboratory by observing all rules themselves. This behavior applies to such duties as wearing goggles and protective clothing, and not working alone in the laboratory. Planning is essential to laboratory safety, and that planning must include what to do in an emergency, as well as how to prevent accidents.

The activities in all *Glencoe* science programs are designed to minimize dangers in the laboratory. Even so, there are no guarantees against accidents. However, careful planning and preparation as well as being aware of hazards can help keep accidents to a minimum. Much information is available on laboratory safety with detailed instructions on planning safe procedures and preventing accidents. However, much of what they present can be summarized in the phrase, "Be prepared!" Know the rules and what common violations occur. Know where emergency equipment is stored and how to use it. Practice good laboratory housekeeping and management by observing these guidelines.

Classroom/Laboratory

1. Store chemicals properly.
   a. Separate chemicals by reaction type.
   b. Label all chemical containers. Include purchase date, special precautions, and expiration date.
   c. Discard outdated chemicals according to appropriate disposal methods.
   d. Do not store chemicals above eye level.
   e. Wood shelving is preferable to metal. All shelving should be firmly attached to walls. Antiroll lips should be placed on all shelves.
   f. Store only those chemicals that you plan to use. Do not stockpile chemicals.
   g. Flammable and toxic chemicals require special storage containers. Do not store more than 500 mL of flammable liquids in the laboratory at one time.
2. Store equipment properly.
   a. Clean and dry all equipment before storing.
   b. Protect electronic equipment and microscopes from dust, humidity, and extreme temperatures.
   c. Label and organize equipment so that it is accessible.
3. Provide adequate work space for students to do investigations.
4. Provide adequate room ventilation.
5. Post safety and evacuation guidelines.
6. Check to ensure that safety equipment is accessible and working properly. Ideally, safety equipment should include fire extinguishers, fire blankets, and eyewash stations.
7. Provide containers for disposing of chemicals, waste products, and biological specimens. Disposal methods must meet local guidelines.
8. Use hot plates for activities requiring a heat source. Be sure the room has an adequate number of electrical outlets, and use only UL-approved extension cords. If laboratory burners are used, a central shutoff valve for the gas supply should be accessible to the teacher. Never use open flames when a flammable solvent is in the same room. Thus, alcohol burners should not be used; alcohol in the presence of fire is a potentially dangerous situation.

First Day of Class/Labs (with students)

1. Distribute and discuss safety rules, first aid guidelines, and safety contract found on page 10 in this *Lab and Safety Skills* booklet.

2. Review safe use of equipment, chemicals, and biological specimens.
3. Review use and location of safety equipment and evacuation guidelines.
4. Discuss safe disposal of materials and laboratory cleanup policy.
5. Discuss proper attitude for working in the laboratory.
6. Document students' understanding of above points.
   a. Have students sign the safety contract and return it.
   b. Administer the safety assessment. Reteach those points that students do not understand.
7. Review the above points with students often during the school year.

Before Each Activity

1. Perform each activity yourself before assigning it to students to determine where students may have trouble.
2. Arrange the lab in such a way that equipment and supplies are clearly labeled and easily accessible. Avoid confusion in the area where solutions and reagents are dispensed.
3. Have available only equipment and supplies needed to complete the assigned activity. This practice helps eliminate the problem of students doing unauthorized experiments.
4. Review the procedure with students. Emphasize cautions within the procedure.
5. Be sure all students know proper procedures to follow if an accident should occur.

During the Activity

1. Make sure the laboratory is clean and free of clutter.
2. Insist that students wear goggles and aprons.
3. Never allow students to work alone.
4. Never allow students to use a cutting device with more than one edge.
5. Shield systems under pressure or vacuum. Use extreme caution if you use a pressure cooker for sterilization purposes. Turn off the heat source and allow the pressure to return to normal before opening the cover.
6. Students should not point the open end of a heated test tube toward anyone.
7. Remove broken or chipped glassware from use immediately. Also clean up any spills immediately. Dilute solutions with water before removing.
8. Be sure all glassware that is to be heated is of a heat-treated type that will not shatter.
9. Remind students that heated glassware looks cool.
10. Prohibit eating and drinking in the lab.

After the Activity

1. Be sure that the laboratory is clean, including all work surfaces and equipment.
2. Be certain that students have disposed of broken glassware and chemicals properly.
3. Be sure all hot plates and burners are off.
4. Insist that each student wash his or her hands when the laboratory work is completed.

NAME _____ DATE _____ CLASS _____

SAFETY SYMBOLS

All *Glencoe* science programs use safety symbols to alert you to possible laboratory dangers. These symbols are explained below. Be sure you understand each symbol before you being an activity.

	DISPOSAL ALERT This symbol appears when care must be taken to dispose of materials properly.		**ANIMAL SAFETY** This symbol appears whenever live animals are studied and the safety of the animals and the students must be ensured.
	BIOLOGICAL HAZARD This symbol appears when there is danger involving bacteria, fungi, or protists.		**RADIOACTIVE SAFETY** This symbol appears when radioactive materials are used.
	OPEN FLAME ALERT This symbol appears when use of an open flame could cause a fire or an explosion.		**CLOTHING PROTECTION SAFETY** This symbol appears when substances used could stain or burn clothing.
	THERMAL SAFETY This symbol appears as a reminder to use caution when handling hot objects.		**FIRE SAFETY** This symbol appears when care should be taken around open flames.
	SHARP OBJECT SAFETY This symbol appears when a danger of cuts or punctures caused by the use of sharp objects exists.		**EXPLOSION SAFETY** This symbol appears when the misuse of chemicals could cause an explosion.
	FUME SAFETY This symbol appears when chemicals or chemical reactions could cause dangerous fumes.		**EYE SAFETY** This symbol appears when a danger to the eyes exists. Safety goggles should be worn when this symbol appears.
	ELECTRICAL SAFETY This symbol appears when care should be taken when using electrical equipment.		**POISON SAFETY** This symbol appears when poisonous substances are used.
	SKIN PROTECTION SAFETY This symbol appears when use of caustic chemicals might irritate the skin or when contact with microorganisms might transmit infection.		**CHEMICAL SAFETY** This symbol appears when chemicals used can cause burns or are poisonous if absorbed through the skin.

SAFETY RULES

1. Always obtain your teacher's permission before beginning an activity.
2. Study the procedure. If you have questions, ask your teacher. Be sure you understand any safety symbols shown on the page.
3. Use the safety equipment provided for you. Goggles and a safety apron should be worn when any activity calls for using chemicals.
4. Always slant test tubes away from yourself and others when heating them.
5. Never eat or drink in the lab, and never use lab glassware as food or drink containers. Never inhale chemicals. Do not taste any substance or draw any material into a tube with your mouth.
6. If you spill any chemical, wash it off immediately with water. Report the spill immediately to your teacher.
7. Know the location and proper use of the fire extinguisher, safety shower, fire blanket, first aid kit, and fire alarm.
8. Keep all materials away from open flames. Tie back long hair and loose clothing.
9. If a fire should break out in the classroom, or if your clothing should catch fire, smother it with the fire blanket or a coat, or get under a safety shower. **NEVER RUN.**
10. Report any accident or injury, no matter how small, to your teacher.

Follow these procedures as you clean up your work area.
1. Turn off the water and gas. Disconnect electrical devices.
2. Return all materials to their proper places.
3. Dispose of chemicals and other materials as directed by your teacher. Place broken glass and solid substances in the proper containers. Never discard materials in the sink.
4. Clean your work area.
5. Wash your hands thoroughly after working in the laboratory.

| **First aid** ||
Injury	Safe response
Burns	Apply cold water. Call your teacher immediately.
Cuts and bruises	Stop any bleeding by applying direct pressure. Cover cuts with a clean dressing. Apply cold compresses to bruises. Call your teacher immediately.
Fainting	Leave the person lying down. Loosen any tight clothing and keep crowds away. Call your teacher immediately.
Foreign matter in eye	Flush with plenty of water. Use eyewash bottle or fountain.
Poisoning	Note the suspected poisoning agent and call your teacher immediately.
Any spills on skin	Flush with large amounts of water or use safety shower. Call your teacher immediately.

SAFETY CONTRACT

I, _____ , have read and understand the safety rules and first aid information listed above. I recognize my responsibility and pledge to observe all safety rules in the science classroom at all times.

_____   _____
signature   *date*

LABORATORY DESIGN AND EQUIPMENT

Sufficient work space is essential in operating an activity-oriented science program. Both a demonstration table and space for students to work in small groups are needed. A portable laboratory demonstration cart is useful if the science classroom lacks a demonstration table and utilities. These carts are available from several scientific suppliers. They usually contain a work space, a sink, a gas torch or burner, water reservoirs, and storage cabinets. See Figure 1. One advantage of the portable demonstration cart is that it can be moved to any location in the classroom and from room to room for science instruction. It can be used in a large room for group instruction.

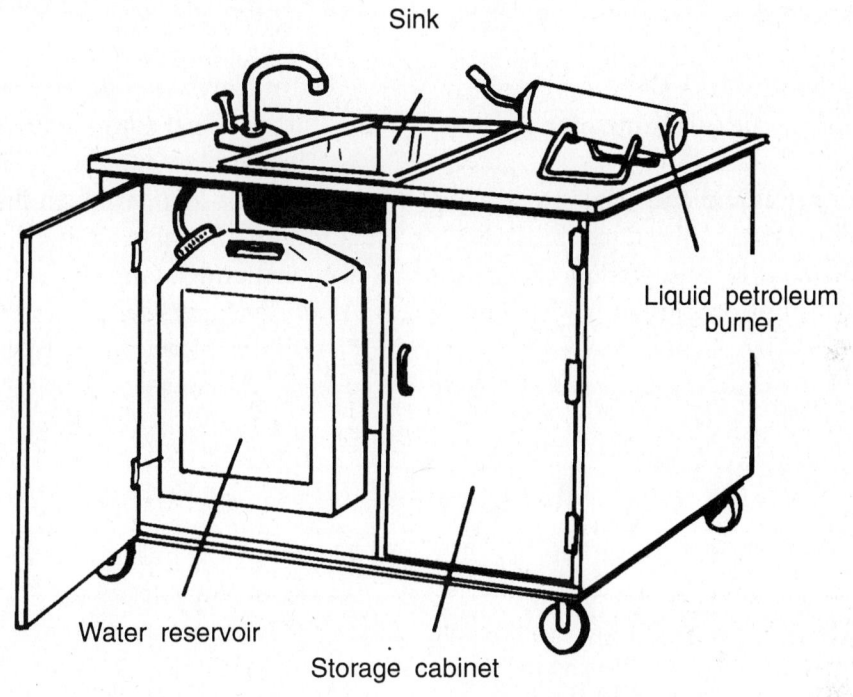

FIGURE 1 — Sink, Liquid petroleum burner, Water reservoir, Storage cabinet

Equipment may be stored in cabinets or in labeled cartons. When equipment is stored in clear containers, the contents can be examined easily without being scattered or damaged. Each container must be clearly labeled as to its contents. If an adequate supply of equipment is available, you may want to organize it in kits. Each kit should contain an equipment checklist as shown here.

Equipment checklist				
Name _____ Date _____				
Quantity	Description	Check in	Check out	Breakage
2	beaker (100 mL)			
2	beaker (250 mL)			
1	beaker (500 mL)			
1	burner			
2	collecting bottle			
2	dropper			

Each team of students would be responsible for filling out and returning this checklist to you at specific intervals, such as once a month or once each semester. These lists will enable you not only to keep track of materials and breakage, but also to assess students' proficiency in keeping their equipment clean and organized. Equipment and specimens that are used only a few times during the year should be labeled and stored in containers in a central location separate from the student kits. Keep a notebook or card file of equipment and materials needed for each activity, as shown below.

Activity 1–2, page 10	
Does air expand or contract when heated?	
Materials	Chemicals
2 balloons	none
2 glass tubes	
2 Erlenmeyer flasks	
2 one-hole rubber stoppers	
mineral oil	
twist ties	
hot plate	

These records will help you determine what materials and equipment are needed ahead of time. They can also help you in ordering materials at the beginning of the school year.

If your supply of laboratory equipment and materials is limited, available materials can often be substituted. For example, different sizes of baby food jars can be substituted for beakers. The regular-sized jar is about the same size as a 100 mL beaker. The junior-sized jar approximates a 250 mL beaker. Use only borosilicate Pyrex glassware for heating materials. Graduated baby bottles (in SI units) or SI measuring cups can be substituted for graduated cylinders.

Storing solutions and chemicals in clearly labeled dropper bottles will reduce spills and waste. Several activities require students to use a few drops of a solution. You may wish to pour the solution into a labeled beaker and provide several droppers for your students.

Remind your students that it is important to use a dropper in only one solution. Dropper bottles are available from drugstores.

Chemical solutions should be prepared prior to the activity and only by the teacher. You may wish to prepare enough stock solution to supply several activities. Your chemical storage areas should be limited to teacher access only. Students should never be allowed to dispense their own chemicals or mix their own solutions from concentrated chemical stock.

Hot plates are recommended for activities requiring a heat source. Be sure your classroom has an adequate number of electric outlets and use only UL-approved extension cords. If laboratory burners are used, be certain that students understand the safe use of burners. Alcohol burners should not be used in the classroom. Alcohol in the presence of fire is a potentially dangerous situation.

EVALUATING ACTIVITY WORK

Evaluation of the activities and of the general outcomes of laboratory work is a difficult task. Pure recognition and recall tests are not usually suitable for evaluating laboratory experience. Evaluation methods that depend on accurate observation, recognition of pertinent data, and ability to reason logically are more suitable for measuring outcomes of laboratory work. This type of evaluation may be done in a variety of ways.
- Periodic checking of student notebooks
- Individual or group conferences
- Requiring student to submit laboratory reports

Laboratory reports should include
- a clearly stated problem.
- a procedure outlined in detail.
- data organized in good form and understandable; may include
  a. labeled diagrams.
  b. labeled and titled graphs.
  c. data tables.
- conclusions that answer the problem based on data obtained in the activity.
- a report that is clear enough to serve as a future review of the material.

The following questions should be answered in evaluating an activity report.
- Is the report written clearly enough so that an uninformed person could read it and know exactly what was being attempted, how it was done, and what conclusions were reached?
- Can the student duplicate the experiment using the report alone as a guide?

Procedures such as identification of an unknown are a common practice in the sciences. Achievement tests designed to assess understanding of course content are an important evaluation technique for laboratory work because of the concern that correct knowledge be obtained through the methods of the laboratory.

Several kinds of tests can be used to evaluate learning in the laboratory.
- Students may be directed to perform a laboratory task in a practical test.
- You can directly observe techniques used, the correctness of procedures, and the results obtained.
- You can prepare an observational checklist based on objectives.

Follow the instructions below to set up a practical test based on the laboratory techniques presented on pages 14 through 20 of this *Lab and Study Skills* booklet. Students should be able to satisfactorily complete this practical test before beginning actual laboratory work. Set up equipment stations in various parts of the classroom. At each station, provide instructions telling students what to do. Use a checklist like the one shown at the end of these instructions for evaluating each student's performance.

Station 1: Lighting a Laboratory Burner
Equipment: laboratory burner, rubber hose, gas outlet, gas lighter, or safety matches
Instructions: Correctly set up and light the burner and adjust the flame.

Station 2: Decanting and Filtering
Equipment: 2 beakers—one containing a mixture of water and sand, stirring rod, filter paper, funnel, ring stand
Instructions: Decant the clear liquid from the residue. Correctly set up the equipment for a filtration procedure.

Station 3: Using the Balance
Equipment: balance, rubber stopper
Instructions: Correctly carry the balance from Station 3 to your desk and back to Station 3. Determine the mass of the rubber stopper.

Station 4: Measuring Temperature
Equipment: thermometer, beaker of water
Instructions: Position the thermometer correctly and determine the temperature of the water in the beaker.

Station 5: Measuring Volume
Equipment: graduated cylinder containing colored water.
Instructions: Determine the volume of water in the graduated cylinder.

Station 6: Identifying Parts of a Microscope
Equipment: microscope, labels
Instructions: Correctly identify the labeled parts of this microscope.

Station 7: Using a Microscope
Equipment: microscope, prepared slide
Instructions: Correctly carry this microscope from Station 7 to your desk and back to Station 7. Place the slide on the stage and bring the slide into sharp focus.

Station 8: Transferring Chemicals from Reagent Bottles
Equipment: reagent bottle containing 50 mL of colored water, 100-mL graduated cylinder, spatula, filter paper, reagent bottle containing salt
Instructions: Transfer the liquid from the reagent bottle to the graduated cylinder. Transfer some of the solid from the reagent bottle to the filter paper.

Station 9: Diluting Acids
Equipment: reagent bottle containing colored water, beaker, stirring rod, jug of distilled water
Instructions: Demonstrate the correct technique for diluting an acid.

Station 10: Inserting Glass Tubing into a Rubber Stopper
Equipment: Glass tubing, glycerol or soapy water, one-hole rubber stopper, cloth towel
Instructions: Insert the glass tubing into the rubber stopper.

Name _____ Date _____
Rating Scale: 1. Student is careless.
2. Student needs to improve.
3. Student is proficient.

Skill	Proficiency
lights burner correctly	1 2 3
adjusts air and gas supply correctly	1 2 3
decants correctly using stirring rod	1 2 3
folds filter paper correctly	1 2 3
carries balance correctly	1 2 3
determines mass accurately	1 2 3
positions thermometer correctly	1 2 3
records accurate temperature reading	1 2 3
uses graduated cylinder correctly	1 2 3
records accurate volume reading	1 2 3
identifies the parts of a microscope	1 2 3
carries microscope correctly	1 2 3
focuses microscope correctly	1 2 3
transfers liquid properly	1 2 3
transfers solid properly	1 2 3
dilutes acid properly	1 2 3
inserts glass tubing correctly	1 2 3

LABORATORY TECHNIQUES

1 Lighting a Laboratory Burner

Most laboratory burners are constructed similarly. There is an inlet for gas and a vent or valve for the adjustment of air, which is mixed with the gas, as shown in Figure 1. For maximum heat, the air-gas mixture must be correct and the object to be heated should be placed just above the pale blue part of the flame.

1. To light the burner, hold a lighted match or a gas lighter next to the barrel of the burner and then turn on the gas.
2. After lighting the burner, adjust the flame from the gas inlet. If the flame rises from the burner or appears to "blow out" after lighting, reduce the supply of gas. Adjust the air vent until a light blue cone appears in the center of the flame. If the flame is yellow, open the air vent.

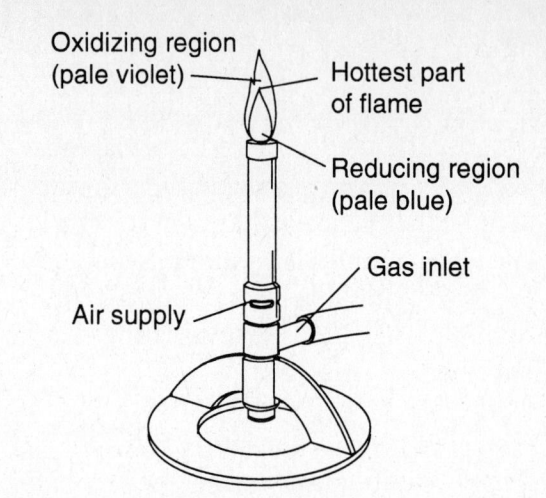

FIGURE 1. Laboratory burner

2 Decanting and Filtering

It's often necessary to separate a precipitate from a liquid. The most common process of separation used in laboratories is filtration.

1. The major portion of the liquid is decanted, or separated from the precipitate, by carefully pouring off the liquid, leaving the solid material. To avoid splashing and to maintain control, the liquid is poured down a stirring rod, as shown in Figure 2.
2. Fold the filter paper as shown in Figure 3.
3. Filter the solution by pouring it through filter paper that catches any remaining precipitates, as shown in Figure 4.
4. Rinse the solid with distilled water to remove any solvent particles. The rinse water should be decanted and filtered.

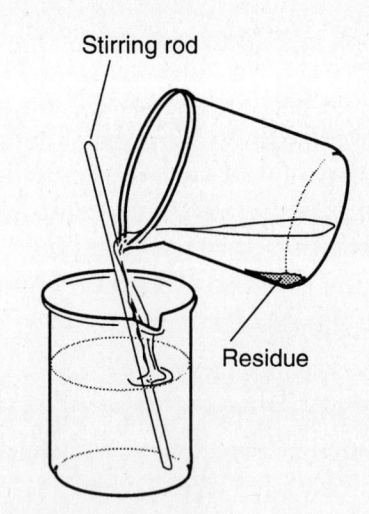

FIGURE 2. Decanting a liquid

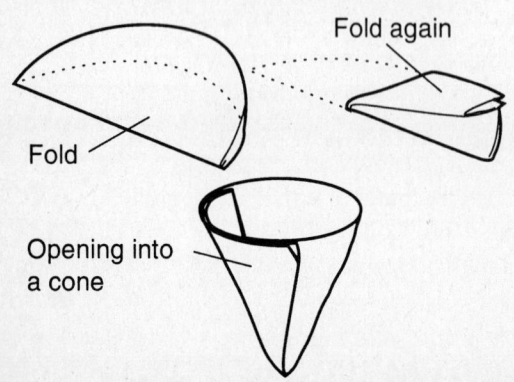

FIGURE 3. Folding a piece of filter paper

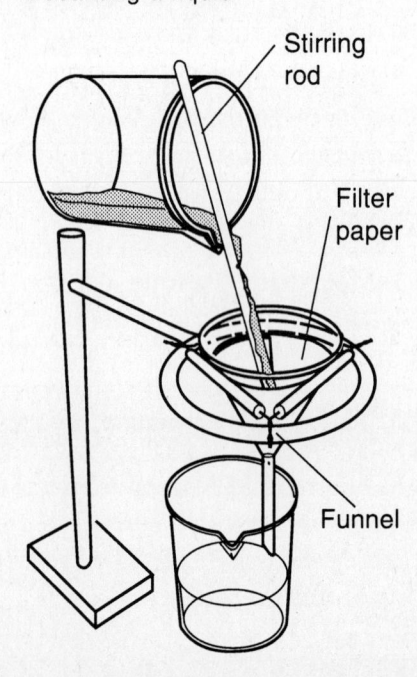

FIGURE 4. Filtering

3 Using the Balance

Although the balance you use may look somewhat different from the balances pictured in Figure 5, all beam balances require similar steps to find an unknown mass.

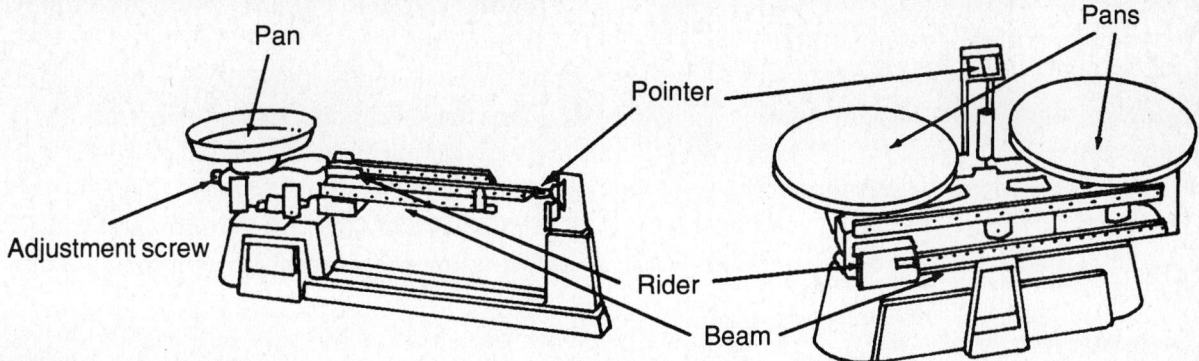

FIGURE 5. Pan balances

Follow these steps when using a beam balance.

1. Slide all riders back to the zero point. Check to see that the pointer swings freely along the scale. The beam should swing an equal distance above and below the zero point. Use the adjustment screw to obtain an equal swing of the beams. You should "zero" the balance each time you use it.
2. Never put a hot object directly on the pan. Air currents developing around the hot object may cause massing errors.
3. Never pour chemicals directly on the balance pan. Dry chemicals should be placed on paper or in a glass container. Liquid chemicals should be massed in glass containers.
4. Place the object to be massed on the pan and move the riders along the beams, beginning with the largest mass first. If the beams are notched, make sure all riders are in a notch before you take a reading. Remember, the swing should be an equal distance above and below the zero point on the scale.
5. The mass of the object will be the sum of the masses indicated on the beams, as shown in Figures 6 and 7. Subtract the mass of the container from the total mass reading, if necessary.

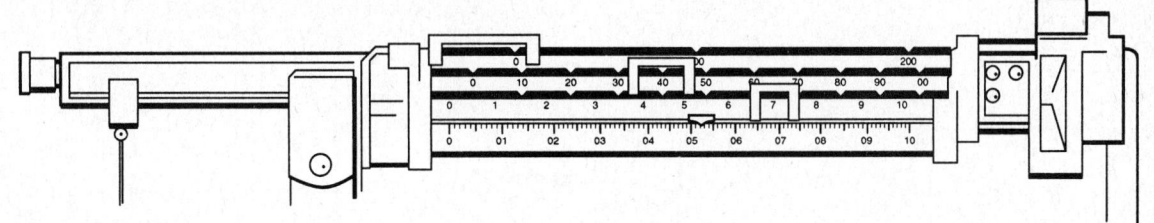

FIGURE 6. The mass of the object would be read as 47.52 grams.

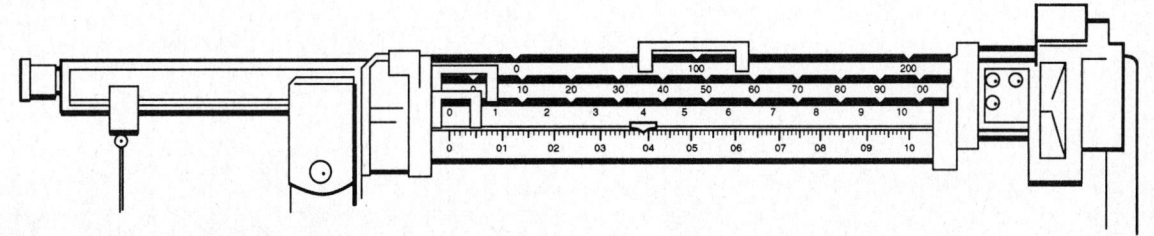

FIGURE 7. The mass of the object would be read as 100.39 grams.

4 Measuring Temperature

1. When the temperature of a liquid is measured with a thermometer, the bulb of the thermometer should be in the center of the liquid. Do not allow the bulb to touch the bottom or sides of the container. When the thermometer is removed from the liquid,

the column in the thermometer will soon show air temperature. For this reason, take temperature readings while the thermometer is in the liquid.
2. When measuring the temperature of hot or boiling liquids, be sure to use a thermometer that is calibrated for high temperatures.
3. Never "shake down" a thermometer to reset it.
4. Never use a thermometer to stir a liquid.

5 Measuring Volume

1. The surface of liquids in glass cylinders is often curved. This curved surface is called the meniscus. Most of the liquids you will measure will have a concave meniscus. View the meniscus along a horizontal line of sight. See Figure 8. Do not try to make a reading looking up or down at the meniscus.
2. Always read a concave meniscus from the low point of the curve. This gives the most precise volume in a glass container.
3. Liquids in many plastic cylinders will not form a meniscus. In these containers, read the volume from the level of the liquid.

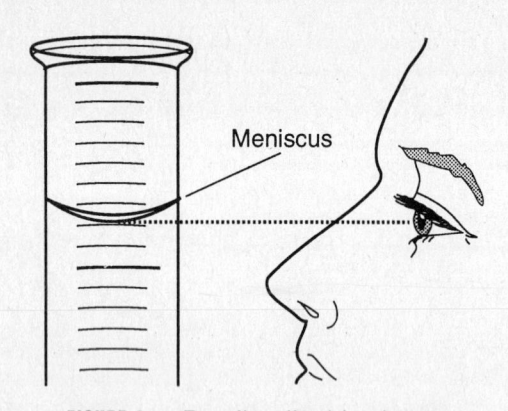

FIGURE 8. Reading liquid volume

6 Transferring Chemicals from Reagent Bottles

CAUTION: *Never touch chemicals with your hands.*

Many chemicals are corrosive and irritating to the skin. Goggles and aprons must be worn when transferring chemicals from one container to another. To avoid contaminating stock chemicals, do not return unused chemicals to the stock bottle.

Solids

1. Solids are generally kept in widemouth bottles. Use a clean spoon or spatula to remove the solid material from its container as shown in Step 1 of Figure 9. Or, rotate the bottle back and forth to shake out the solid.
2. Place the solid material on a piece of creased, waxed paper and add the solid very carefully to your container (Step 2 of Figure 9). Transfer the solid to a test tube by folding the paper as shown in Step 3 of Figure 9.
3. If the solid is to be massed, remember to use waxed paper or a container. Do not place the solid directly on the balance pan.

Step 1

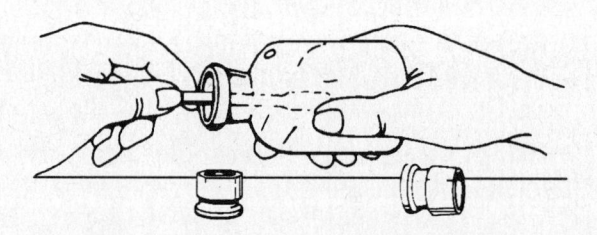

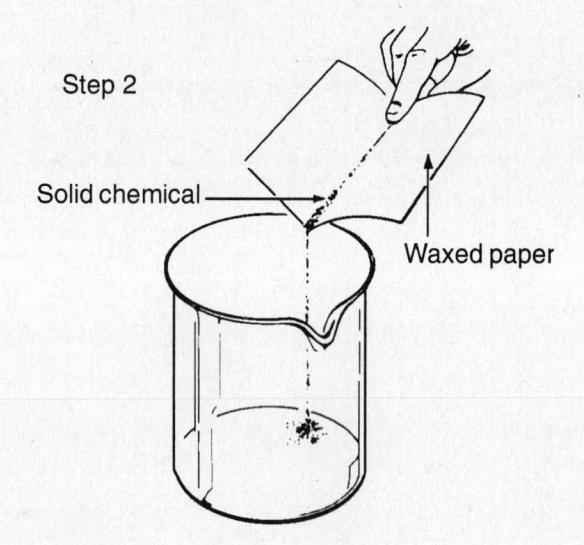

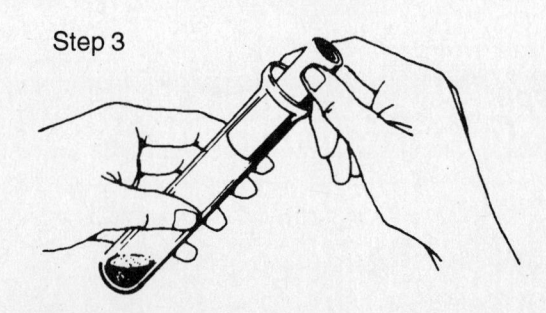

Step 2

Step 3

FIGURE 9. Transferring a solid

Liquids

1. Grasp the stopper between your fingers as shown in Figure 10 and remove the stopper from the reagent bottle. Do not put the stopper on the table; keep it between your fingers.
2. Wearing goggles, hold the test tube or graduated cylinder at eye level and pour the liquid slowly until the desired volume has been transferred. Read the volume as described above.
3. Replace the stopper in the reagent bottle. If any liquid runs down the outside of the bottle, rinse it with water before returning it to the shelf. Wipe the bottle with a damp paper towel if the liquid is an acid.

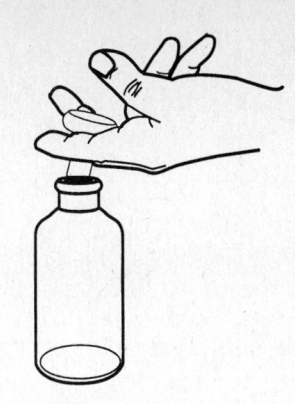

FIGURE 10. Removing a stopper from a reagent bottle

7 Working with Chemicals

1. When smelling a substance, use a fanning motion to direct the vapor toward you. Never smell a substance directly. The proper technique is shown in Figure 11.
2. Always point the mouth of the test tube away from yourself and others when you heat the test tube. Move the tube constantly for even heating. See Figure 12.
3. Diluting acids
   a. The acid is added to the water, and never the reverse.
   b. The acid should be poured slowly down the stirring rod and the solution continually stirred as shown in Figure 13, Diluting an acid produces heat. Therefore, it is important to add the acid slowly and to stir the solution.

FIGURE 11. Smelling a substance

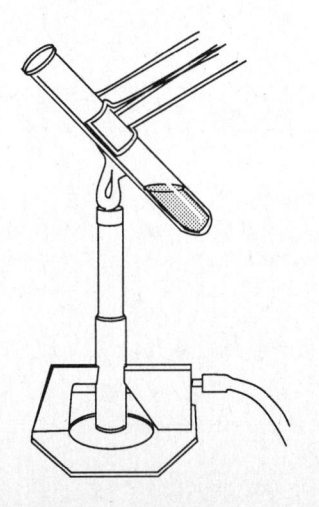

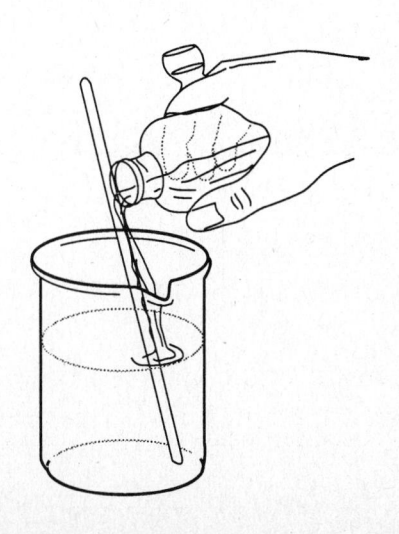

FIGURE 12. Heating a test tube

FIGURE 13. Diluting an acid

8 Inserting Glass Tubing into a Rubber Stopper

1. Begin by lubricating the tip of the glass tubing and the rubber stopper with soapy water, glycerol, or some other suitable substance.
2. Protect your hands with a cloth towel. Be sure that the ends of the tubing are directed away from the palms of your hands. Never force the tubing into the stopper. Ease it in with a gentle twisting motion.
   CAUTION: *Excessive hand pressure on the tubing will cause it to break. Sever injury can occur.*
3. The end of the glass tubing should protrude from the stopper.

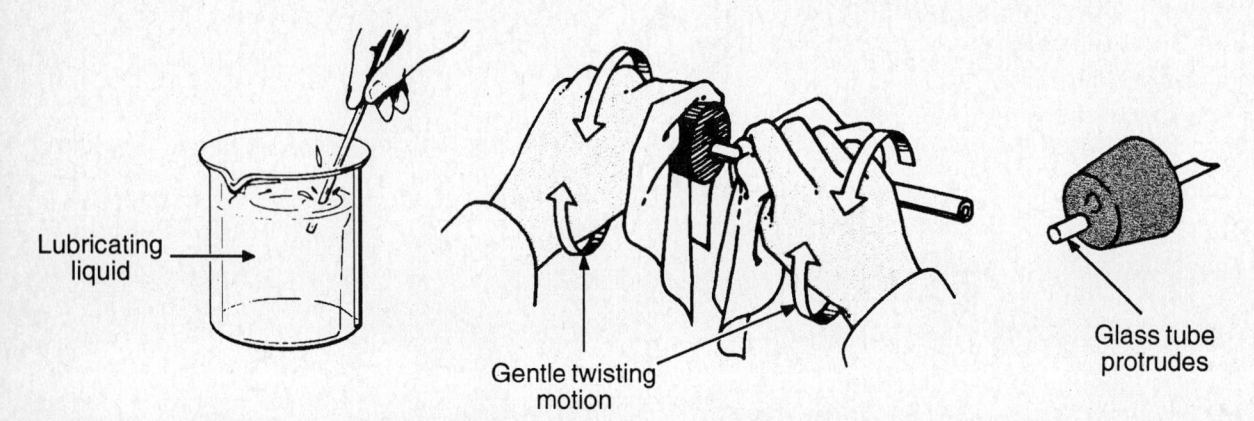

FIGURE 14. Inserting glass tubing into a rubber stopper

9 Microchemistry

Microchemistry uses smaller amounts of chemicals than do other chemistry methods. The hazards of glass have been minimized by the use of plastic labware and hot water provides heat rather than open flames or burners.

1. Reactions occur in a plastic tray called a microplate. The tray has shallow wells arranged in Rows (running across) and Columns (running up and down). These wells are used instead of test tubes, flasks and beakers. Some microplates have 96 wells, arranged as in Rows A–D in Figure 15. Other microplates have 24 larger wells, as shown in the bottom two rows of Figure 15.

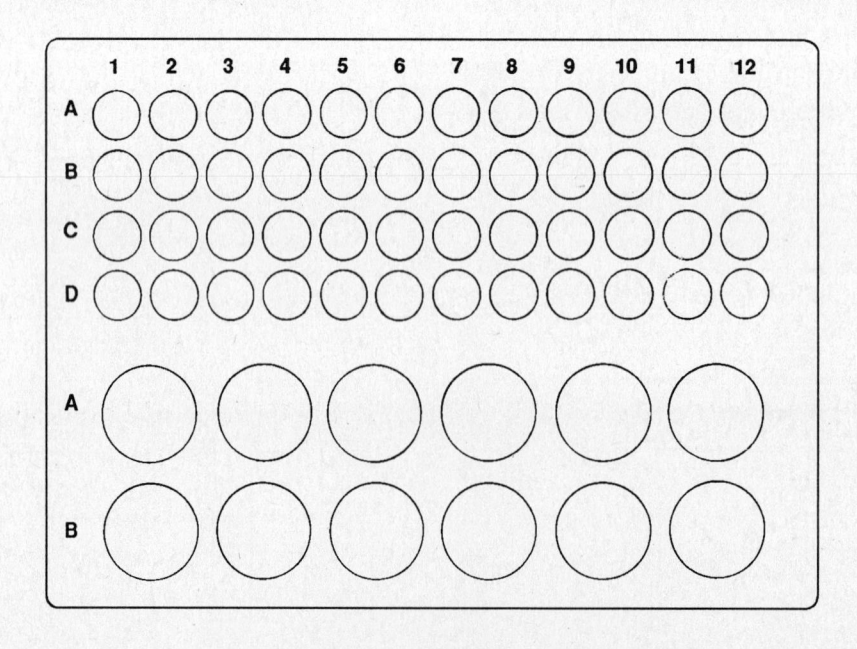

FIGURE 15. Microplate

2. Liquids are transferred in Microchemistry using a soft, very flexible plastic pipet. See Figure 16. The stem of the pipet can be stretched into a thin tube. If the stem is stretched and then cut with scissors (Figure 17), the small tip will deliver a tiny drop of chemical. You may also use a pipet called a Microtip pipet which has been pre-stretched at the factory. It is not necessary to stretch a Microtip pipet. The plastic pipet can be used over and over again simply by rinsing the stem and bulb between chemicals.

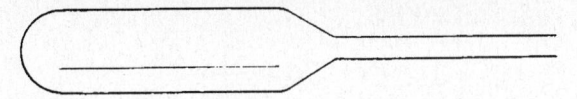

FIGURE 16. Plastic pipet

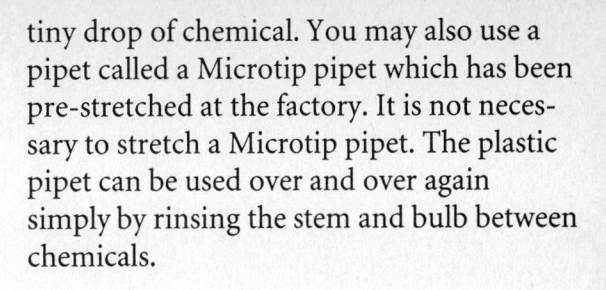

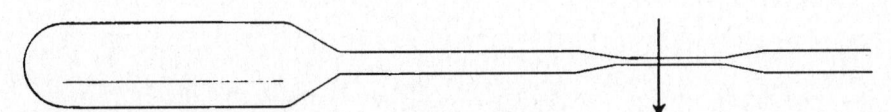

FIGURE 17. Cutting a pipet

10 Identifying Parts of a Microscope

Most microscopes have the parts shown in Figure 18.
  a. Eyepiece
  b. Body tube
  c. Revolving nosepiece
  d. Low-power objective lens
  e. High-power objective lens
  f. Stage
  g. Stage clips
  h. Base
  i. Mirror
  j. Diaphragm
  k. Arm
  l. Fine adjustment knob
  m. Coarse adjustment knob

FIGURE 18. Microscope

11 Using a Microscope

Microscopes should be stored in a safe place where storage and retrieval can be supervised. If not kept in a cabinet, a microscope should be protected from dust with a cloth or plastic cover.

1. Always carry a microscope upright with two hands, one hand holding the arm and one hand supporting the base.
2. Store the microscope with the low-power objective in position.
3. Always bring a specimen into focus with the low-power objective first.
4. Never use the coarse adjustment to focus the high-power objective.
5. When using the coarse adjustment to lower the low-power objective, always look at the microscope from the side. If you look through the eyepiece, you may accidentally force the objective into the coverslip.
6. Do not allow direct sunlight to shine on the mirror and reflect up into the eye.
7. Clean lenses only with lens paper. Moisten the lens paper with a drop of water or alcohol if the lens does not wipe clean with dry lens paper.
8. Be careful when using coverslips and microscope slides because they may crack or shatter when dropped.

12 Using Preserved and Live Animals

1. Use extreme caution when dissecting preserved specimens. Dissecting tools are very sharp. Always use a dissecting pan to support your specimen. Never hand-hold a specimen to dissect it.
2. Many laboratories contain live animals. Always wear heavy gloves when handling animals. If you are bitten, report the bite at once. Do not destroy the animal that has bitten you. Instead, call the local department of health for further instructions.

13 Testing for the Hardness of a Mineral

Hardness is the resistance of a mineral to being scratched. When testing for hardness, one mineral is harder than another if the first mineral can scratch the second.

1. Test a mineral for hardness by scratching the mineral against a mineral with a known hardness. See Figure 19.
2. Reverse the minerals and scratch again to confirm your results. If the minerals are nearly the same hardness, a scratch may be difficult to see.
3. Use a set of Mohs' minerals to determine the hardness of unknown minerals. If Mohs' minerals are not available, you may substitute common objects. If a mineral can be scratched by a fingernail, it has a hardness of about 2; a copper penny, a hardness of 3; a nail, a hardness of 5 to 6; a knife blade, a hardness of 6; and a piece of glass, a hardness of 6 to 7.

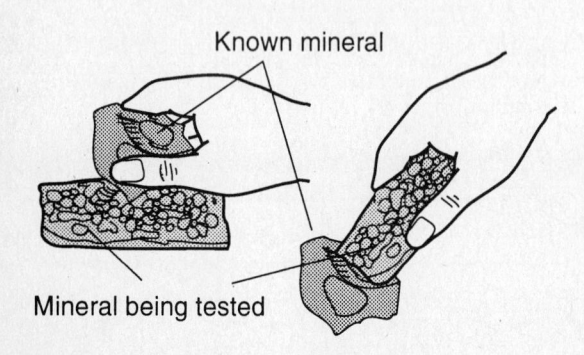

FIGURE 19. Testing for hardness

14 Testing for Streak

1. Streak is the color of the powdered mineral that is left when the mineral is drawn across a nonglazed porcelain plate called a streak plate. Figure 20.
2. The streak is often not the same color as the mineral sample.

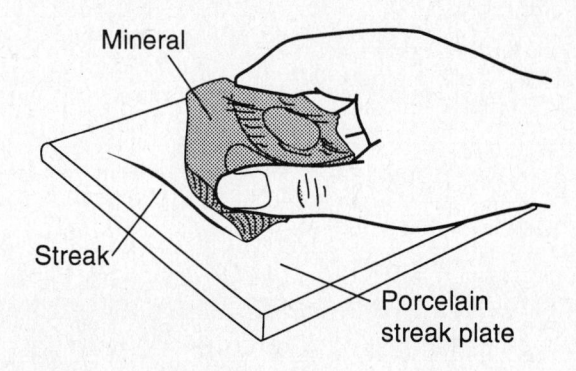

FIGURE 20. Testing for streak

15 Testing for Magnetism

Very few minerals are magnetic.

1. To test a mineral for magnetism, break it into small pieces and test the pieces with a magnet as shown in Figure 21.
2. Some minerals become magnetic only if they are heated. Hold a small piece of the mineral in a flame with tongs. Be sure to wear goggles and use a thermal mitt.
3. Repeat the test with the heated mineral and the magnet.

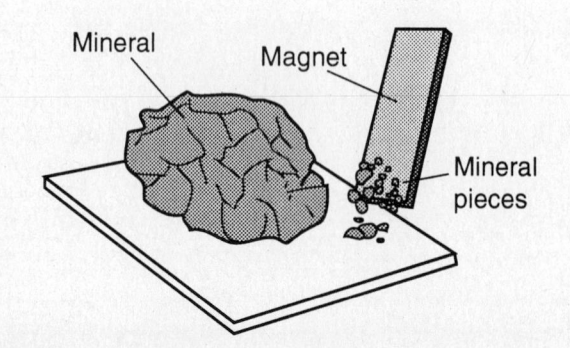

FIGURE 21. Testing for magnetism

16 Testing for Cleavage and Fracture

A mineral has cleavage if it breaks under stress to form smooth, flat reflective faces. Minerals may cleave along different planes that form parallel cleavage surfaces.

1. A test for cleavage is shown in Figure 22.
2. If the mineral breaks or does not break parallel to the chisel, turn the mineral and repeat the process until you have checked for cleavage in all directions.
3. A mineral that does not cleave is said to fracture or break along irregular surfaces.

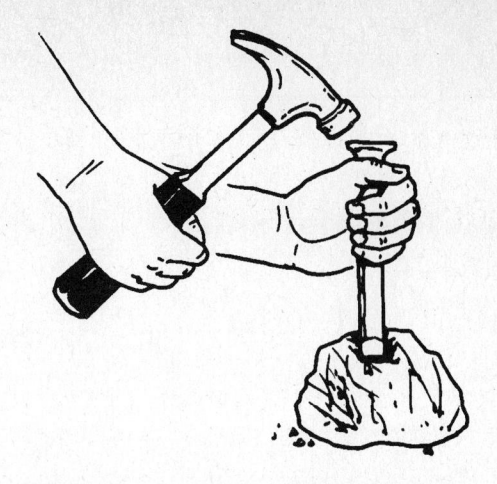

FIGURE 22. Testing for mineral cleavage

NAME _____ DATE _____ CLASS _____

LABORATORY SKILLS ASSESSMENT
Scientific Problem Solving

Lab and Safety Skills

Scientific Methods

All scientists have special interests. Scientists are interested in the world around them. This curiosity leads them to investigate things and events. Scientists use their senses to observe as they investigate. They use many methods in seeking answers to problems. Scientists use scientific problem solving. One scientific problem-solving technique has six steps:

1. State the problem.
2. Gather information about the problem.
3. Form a hypothesis.
4. Test the hypothesis.
5. Accept or reject the hypothesis.
6. Do something with the results.

Using Scientific Methods

Scientists observed that white mice that were fed seeds appeared to grow more than mice given leafy green and yellow vegetables. The scientists hypothesized that the protein in the seed was responsible for the growth. They designed an experiment to test this hypothesis. They divided 200 mice of the same age, size, health, and sex into two groups of 100 mice each. The mice were kept under identical conditions for fourteen days. One group was given a diet low in protein. The other group was given a normal protein diet. The mass of each mouse was recorded daily for fourteen days.

1. Which group of mice served as a control? _____

2. What was the variable? _____

3. What effect of the protein diet was tested? _____
_____

4. What other effects of a protein diet could have been tested? _____
_____

5. Why were larger numbers of mice used in this experiment? _____
_____

6. If the results of the experiment did not show a marked change between the two groups, what should the scientists do next? _____

7. What are the parts of an experiment? _____
_____

LABORATORY SKILLS ASSESSMENT
Scientific Problem Solving

Lab and Safety Skills

Tables and Graphs
A. Making Tables

As you study science, you will find that often information is presented in a table. You will organize data and observations in tables. You will be asked to use information given in tables. Knowing how to make and use tables is an important skill.

Tables have a title, rows, columns, and heads. The title is found at the top of the table. The title tells you what information is contained in the table. Columns are the sections that run up and down. At the top of each column is a head that tells you what information is in the column. Rows are sections that run from one side to another on the table.

Use the information in this paragraph to complete the table below. When you complete a table, you must record the information in the proper column and row. There are 5000 species in Phylum Porifera, 11 000 species in Phylum Coelenterata, 26 000 species in the three phyla of worms, 80 000 species in Phylum Mollusca, 826 000 species in Phylum Arthropoda, 47 000 species in Phylum Chordata, and 5000 species in Phylum Echinodermata.

Number of animal species	
Phylum	Number

NAME _____ DATE _____ CLASS _____

B. Making Graphs

A graph is a picture that shows data in a way that helps you understand the information. There are several different types of graphs. Line graphs show data plotted as points that are connected by a line. Line graphs often are used to show change. The graph in Figure 1 shows the increase in world population over time.

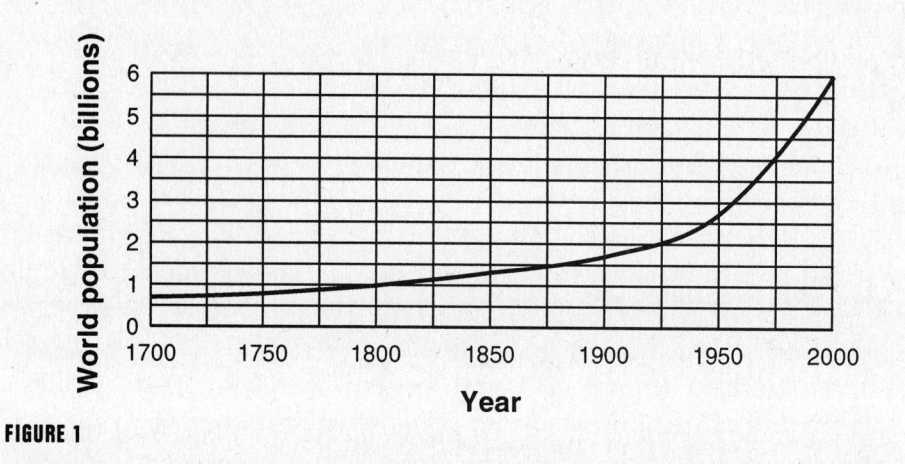

FIGURE 1

A bar graph always has two axes, a horizontal and a vertical, and these axes are labeled and divided. On the bar graph in Figure 2, the vertical axis is labeled and divided into ten thousands. The horizontal axis is labeled "Phylum" and divided by different animal phyla. Indicate the number of species in each phylum given by shading in bars on the graph. The column for Phylum Porifera has been done for you.

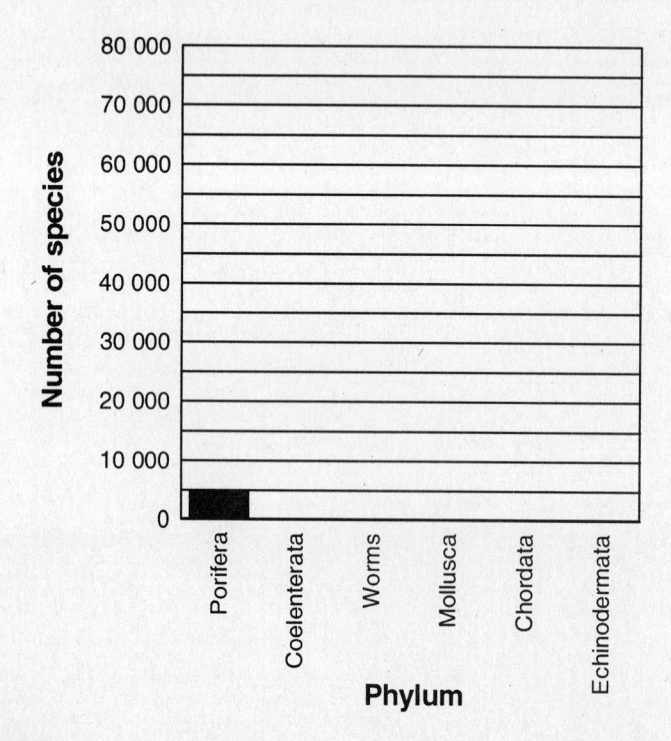

FIGURE 2

NAME _____ DATE _____ CLASS _____

Pie graphs are circular graphs that show how each part is related to the total. Use the data from the table below to make a pie graph. Calculate the fractional amount of time for each of the sections. The total circle will represent 4.5 billion years. To find the fraction for each section, divide the time length of that section by 4 500 000 000. Then multiply the fraction by 360 degrees to determine the angle for the section. The calculation for the Cenozoic Era is given.

65 000 000 years ÷ by 4 500 000 000 years = 0.0144 × 360° = 5.18°

Round the final answer to the nearest whole degree. Use this procedure to complete the table. Then, use a protractor and the data from the table to construct a circular graph in Figure 3. Label all parts of the graph.

Era	Number of years	Number of degrees
Cenozoic	65 000 000	5
Mesozoic	160 000 000	
Paleozoic	345 000 000	
Precambrian	3 930 000 000	

Use a protractor and the data to construct a circular graph in Figure 3. Label all parts of the graph.

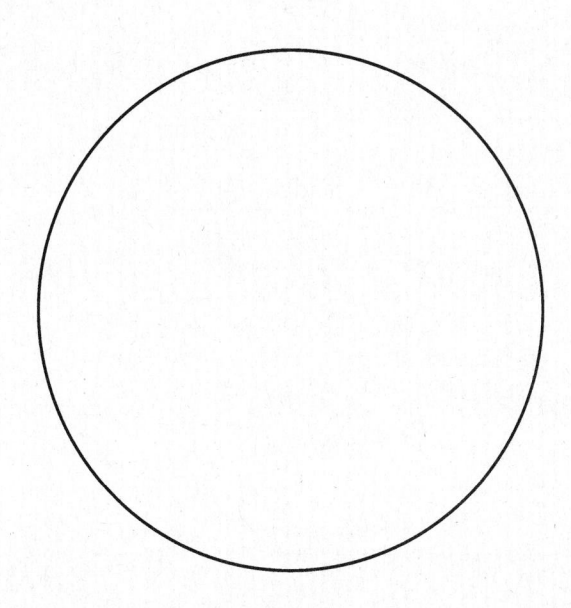

FIGURE 3

NAME _____ DATE _____ CLASS _____

LABORATORY SKILLS ASSESSMENT
Safety in the Science Classroom

Lab and Safety Skills

A. Developing Safety Policies

Several statements are printed in Column I concerning students' activities and attitudes in the laboratory. Think about each statement and formulate a safety rule or procedure related to each statement. Write one or two clear, concise sentences in Column II that can serve as a safety policy for your science classroom. Use Appendix C in your textbook as a reference.

Column I
1. Peter says that his teacher is solely responsible for preventing laboratory accidents.

2. Keshia started the lab activity before reading it through completely.

3. Ricardo decided to do a lab activity that he read in a library book before the teacher came into the classroom.

4. Stephanie says that the lab aprons are unattractive and that the safety goggles mess up her hair. She refuses to wear them.

Column II
1. _____
   _____
   _____

2. _____
   _____
   _____

3. _____
   _____
   _____

4. _____
   _____
   _____

B. Using Safety Devices Correctly

Describe the location and purpose of having each of the following devices in your science laboratory. Write your answers in the spaces provided. (Locations will vary according to your laboratory design.)

1. fire blanket _____
   _____

2. CO_2 fire extinguisher _____
   _____

3. goggles _____

4. eyewash station or fountain _____

5. safety hood or vent _____
   _____

NAME _____ DATE _____ CLASS _____

C. Identifying Improper Safety Practices

The pictures below show students performing laboratory activities incorrectly. Study each picture and write in the space provided all improper laboratory techniques that are illustrated. Be prepared to explain why it is important to follow each safety procedure.

1. _____

_____

2. _____

_____

_____

Copyright © Glencoe Division of Macmillan/McGraw-Hill

27

NAME _____ DATE _____ CLASS _____

Explain what is wrong in each of the following diagrams and tell how to correct it.

3. Error _____

_____

Correction _____

_____

_____

_____

_____

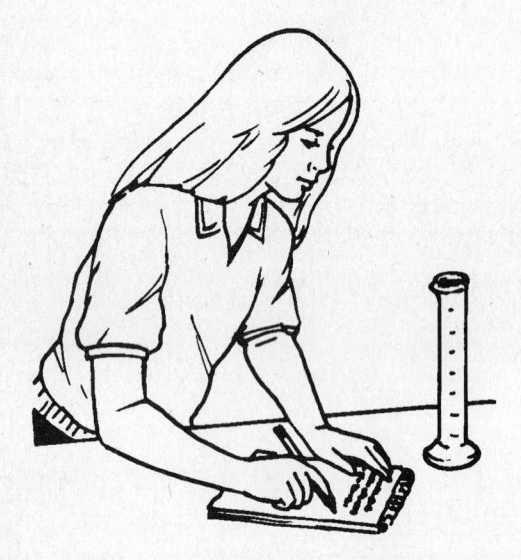

4. Error _____

_____

Correction _____

_____

_____

_____

_____

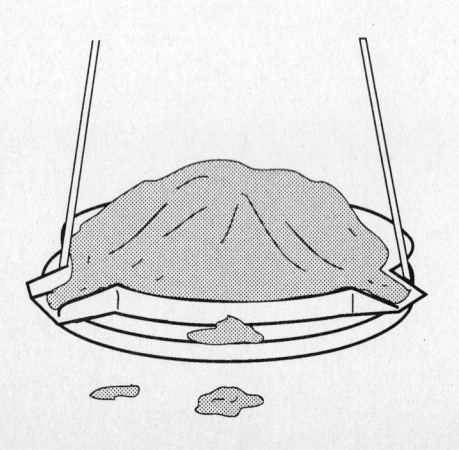

5. Error _____

_____

_____

Correction _____

_____

_____

_____

28   Copyright © Glencoe Division of Macmillan/McGraw-Hill

NAME _____ DATE _____ CLASS _____

6. Error _____
_____
_____

Correction _____
_____
_____
_____

7. Error _____
_____
_____

Correction _____
_____
_____
_____
_____

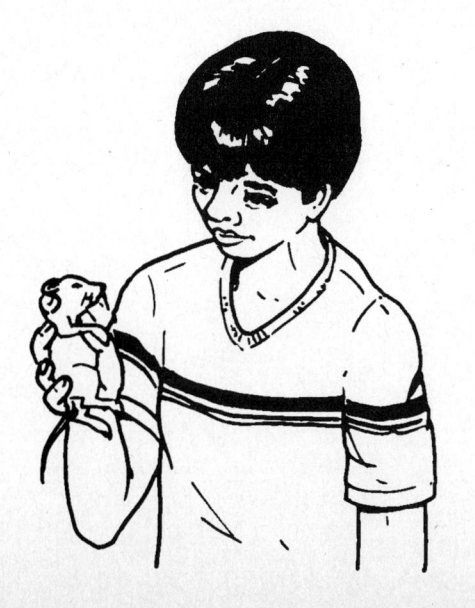

8. Error _____
_____
_____

Correction _____
_____
_____
_____
_____

NAME _____ DATE _____ CLASS _____

LABORATORY SKILLS ASSESSMENT
Using a Microscope

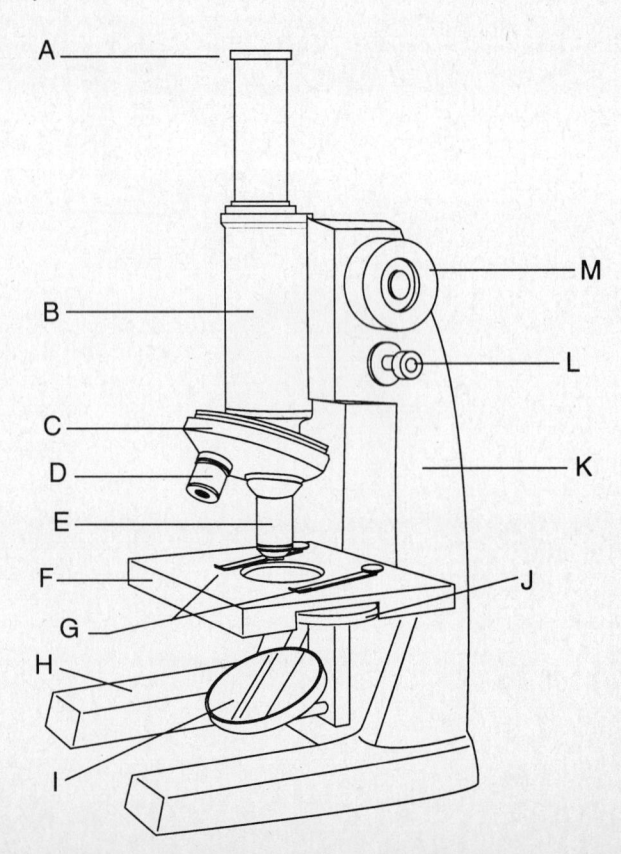

Lab and Safety Skills

Microscopes are optical instruments, much like a pair of glasses. The purpose of the microscope is to help you see things that are very small. The microscope can introduce you to organisms that you would otherwise not see. Because it is a delicate instrument, a microscope requires careful handling.

Show that you are aware of how to handle a microscope correctly by completing the following statements.

1. Always carry a microscope with one hand on the _____ and the other hand on the _____ .

2. A microscope should be stored with the _____ objective in place.

3. Always bring a specimen into focus using the _____ objective.

4. Never use the _____ adjustment to focus the high-power objective.

5. Do not allow direct sunlight to fall on the _____ .

6. Use only the _____ adjustment when focusing with the high-power objective.

7. Lenses should be cleaned with _____ .

Identify the parts of the microscope on the lines to the left.

a. _____
b. _____
c. _____
d. _____
e. _____
f. _____
g. _____
h. _____
i. _____
j. _____
k. _____
l. _____
m. _____

30 Copyright © Glencoe Division of Macmillan/McGraw-Hill

NAME _____ DATE _____ CLASS _____

LABORATORY SKILLS ASSESSMENT
Measuring with a Microscope

Lab and
Safety Skills

Scientists use microscopes to study things that are too small to be seen with the unaided eye. The size of microscopic organisms can be estimated using the method given here. The approximate size of the field of view seen under low power can be determined by actual measurement. A transparent millimeter ruler is placed across the field of view as shown in Figure 1.

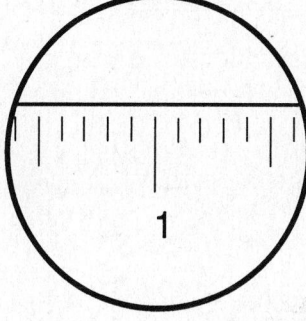

FIGURE 1

1. How many millimeters wide is the field of view shown? _____

   Objects examined under the microscope are quite small. Therefore, it is often convenient to use units of length smaller than the millimeter for microscope measurement. One unit often used is the micron (μ). There are 1000 microns in one millimeter.

2. How many microns wide is the low-power field shown in Figure 1? _____

   To measure the diameter of the high-power field of view, follow these steps. First divide the magnification number of the high-power objective by the magnification number of the low-power objective. Then divide the diameter of the low-power field of view by this number. For example, assume that a low-power objective is 10×, and the higher power objective is 40×. On this microscope, the diameter of the low-power field of view is 2000 microns.

3. What is the width of the high-power field of view? _____

4. What is the width of the organisms shown below? These organisms were viewed under high power.

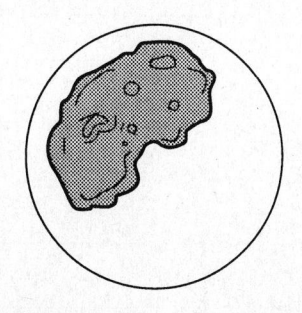

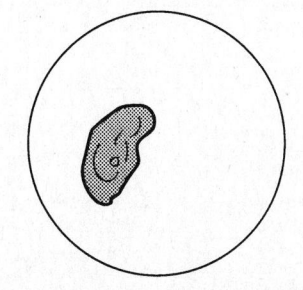

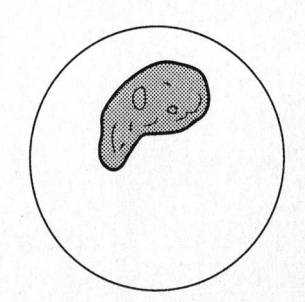

FIGURE 2

NAME _____ DATE _____ CLASS _____

LABORATORY SKILLS ASSESSMENT
Using Laboratory Measuring Devices

Lab and Safety Skills

Balance: Determining Mass

1. What mass is shown on each of these balances?

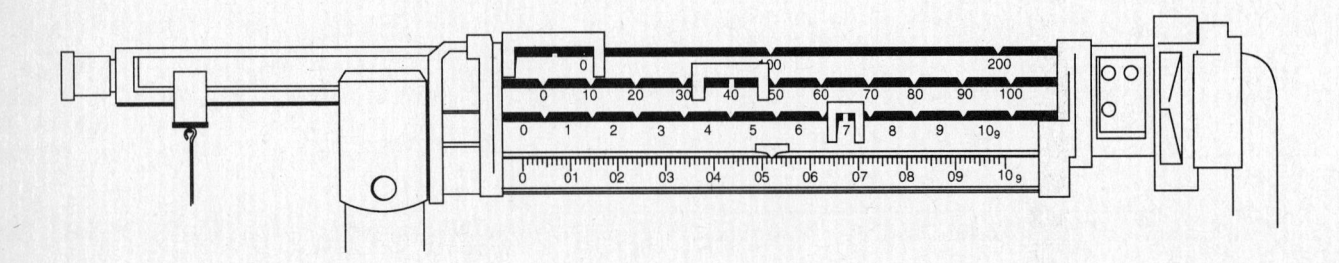

a. The mass of the object would be read as _____ .

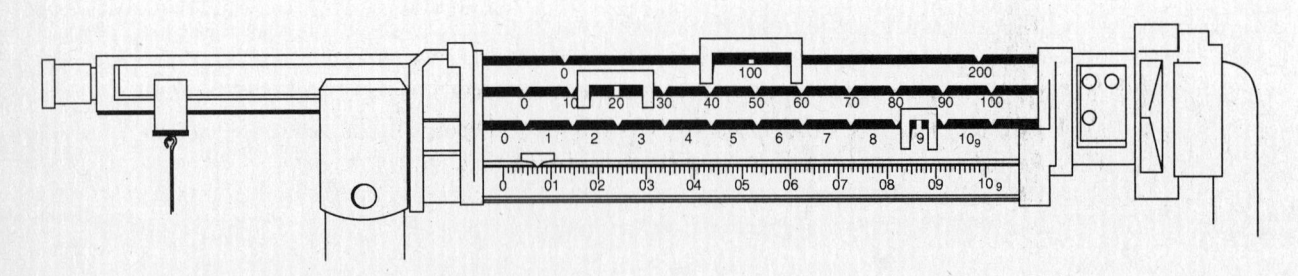

b. The mass of the object would be read as _____ .

Metric Ruler: Determining Length

2. What lengths are indicated on this ruler?

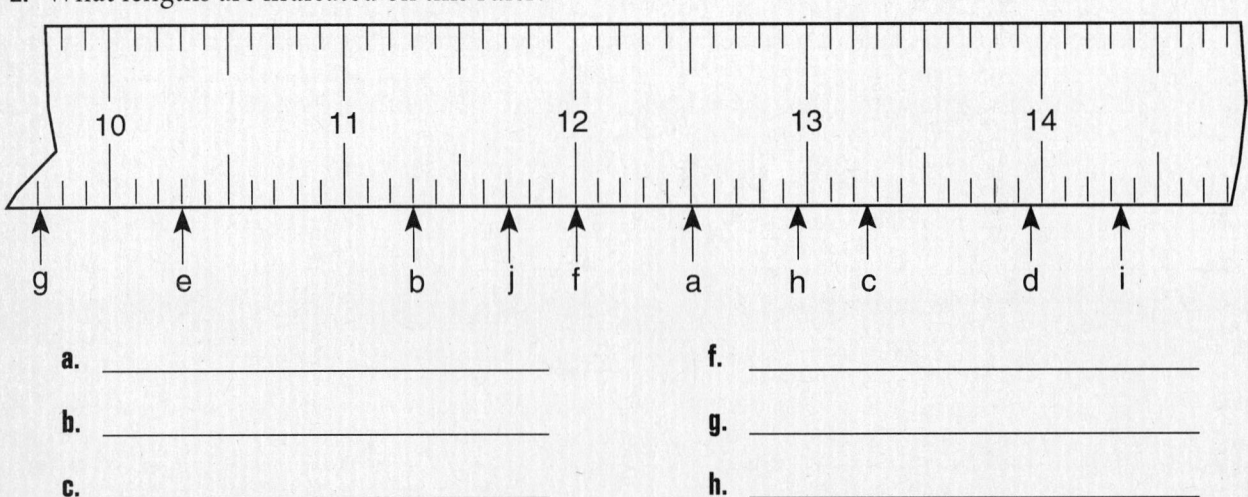

a. _____   f. _____

b. _____   g. _____

c. _____   h. _____

d. _____   i. _____

e. _____   j. _____

32 Copyright © Glencoe Division of Macmillan/McGraw-Hill

NAME _____ DATE _____ CLASS _____

Graduated Cylinder: Measuring Liquid Volume

3. What volume is indicated on each of these graduated cylinders?

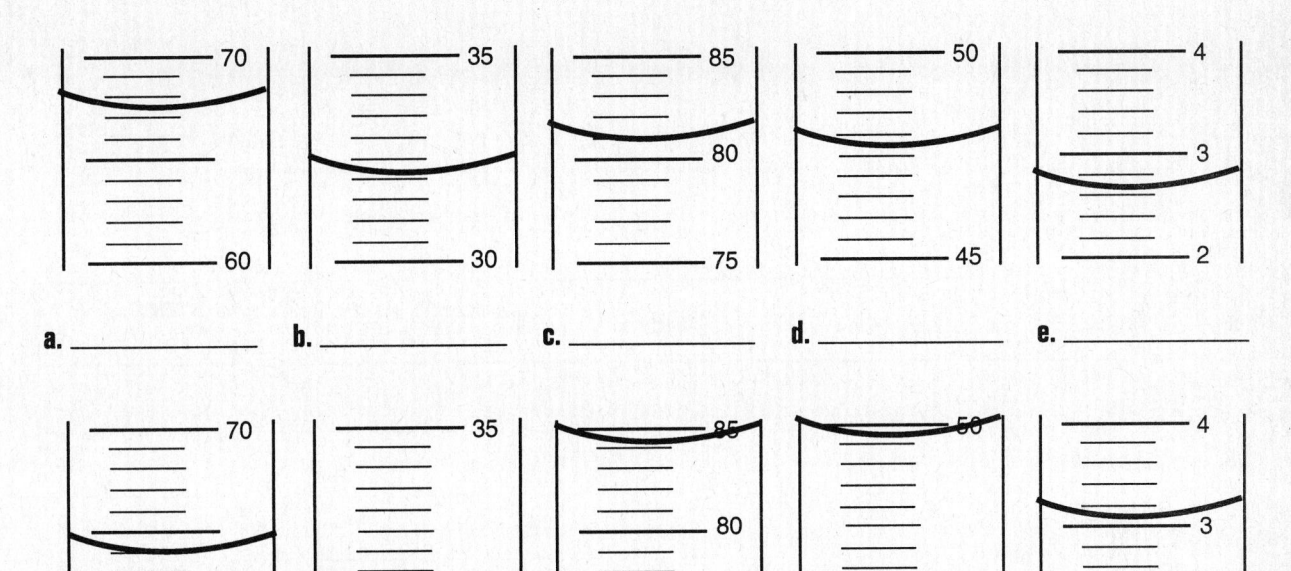

a. _____ b. _____ c. _____ d. _____ e. _____

f. _____ g. _____ h. _____ i. _____ j. _____

Thermometer: Measuring Temperature

4. What temperature is indicated on each of these thermometers?

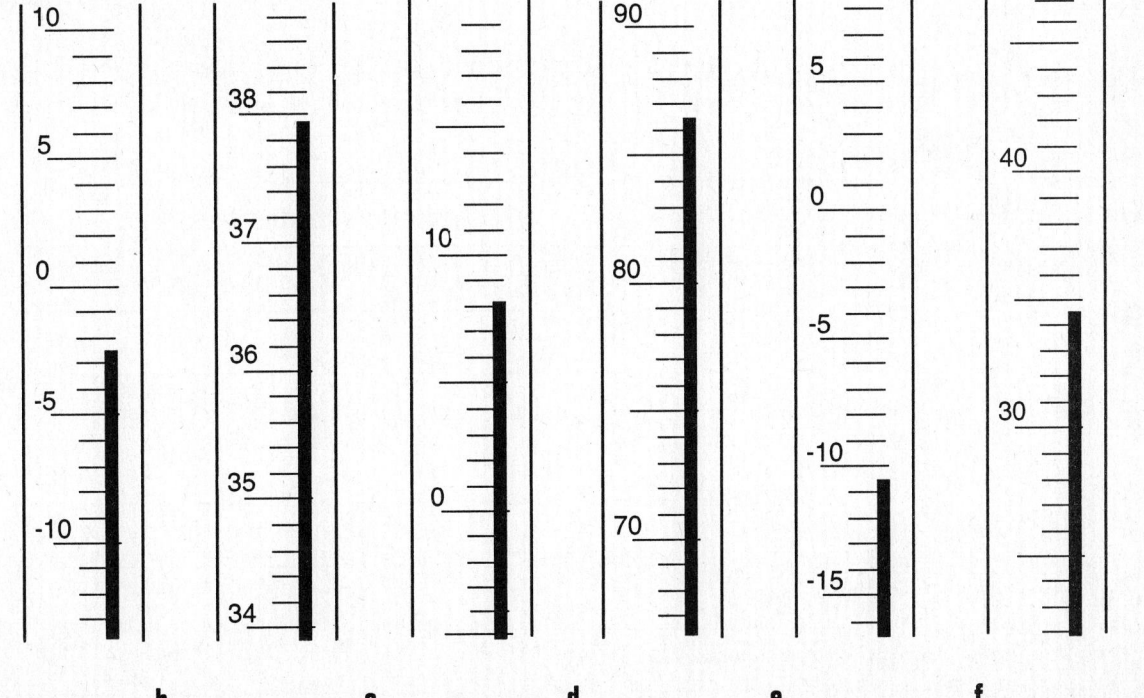

a. _____ b. _____ c. _____ d. _____ e. _____ f. _____

Copyright © Glencoe Division of Macmillan/McGraw-Hill

NAME _____ DATE _____ CLASS _____

LABORATORY SKILLS ASSESSMENT
Using a Metric Ruler and Protractor

Lab and Safety Skills

A. Calculating Surface Area and Volume Using Metric Measurements

Your teacher will give you several objects including those items listed in Table 1. Using your metric ruler, make the required measurements and complete the table. Use your data to calculate surface area or volume for each item.

TABLE 1

Object	Length	Width	Height	Surface area/volume
Index card				
Microscope slide				
Petri dish				
Chalk				
Desk top				

B. Metric Scale Conversions

Using your metric ruler and the map on page 35, measure the distances between the points indicated in Table 2. Record your answers in Table 2. Complete the table using the scale 10 cm = 1 km.

TABLE 2

How far is it from the _____ ?	Metric measurement mm	Metric measurement cm	Actual distance km
Softball field to the lake			
Meadow to the picnic area			
Hickory grove to the lake			
Pine forest to the picnic area			
Softball field to the hickory grove			
Softball field to the lake through the meadow			

Copyright © Glencoe Division of Macmillan/McGraw-Hill

NAME _____ DATE _____ CLASS _____

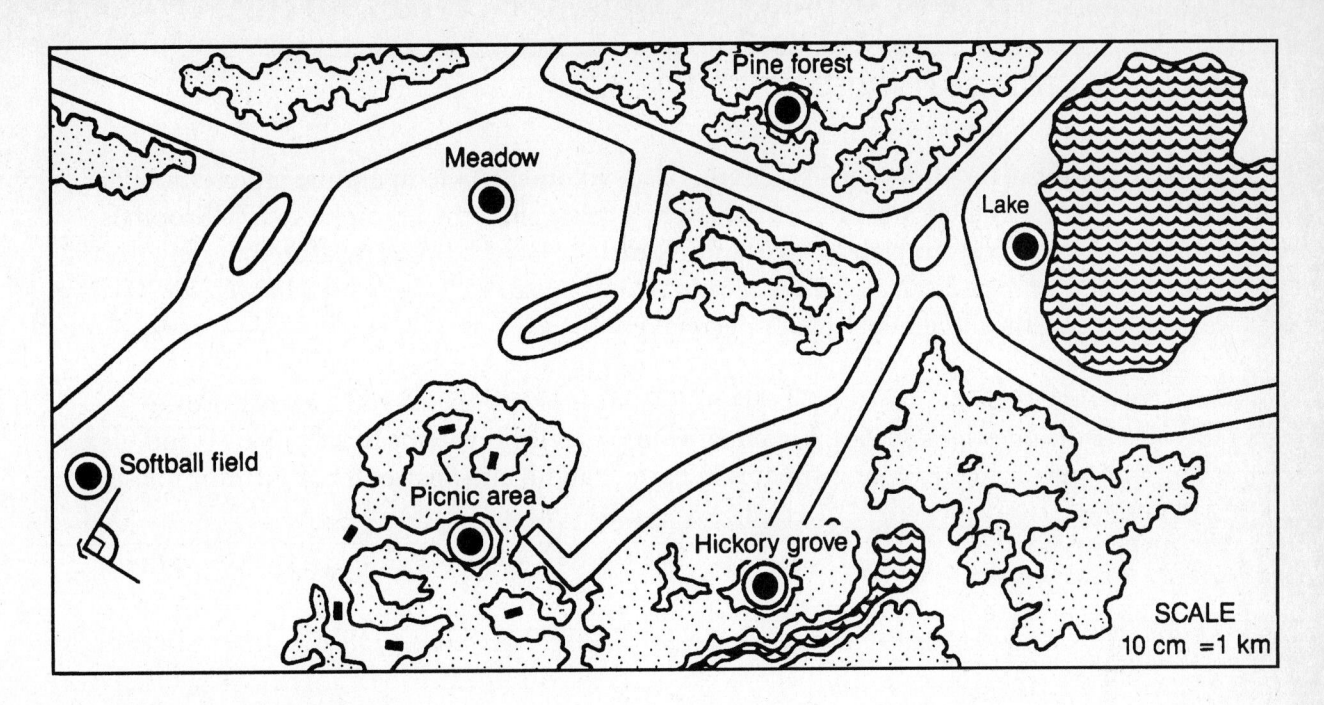

C. Measuring Angles

1. Using your protractor, measure the size of the angles below. Record your answers on the lines after the letters.

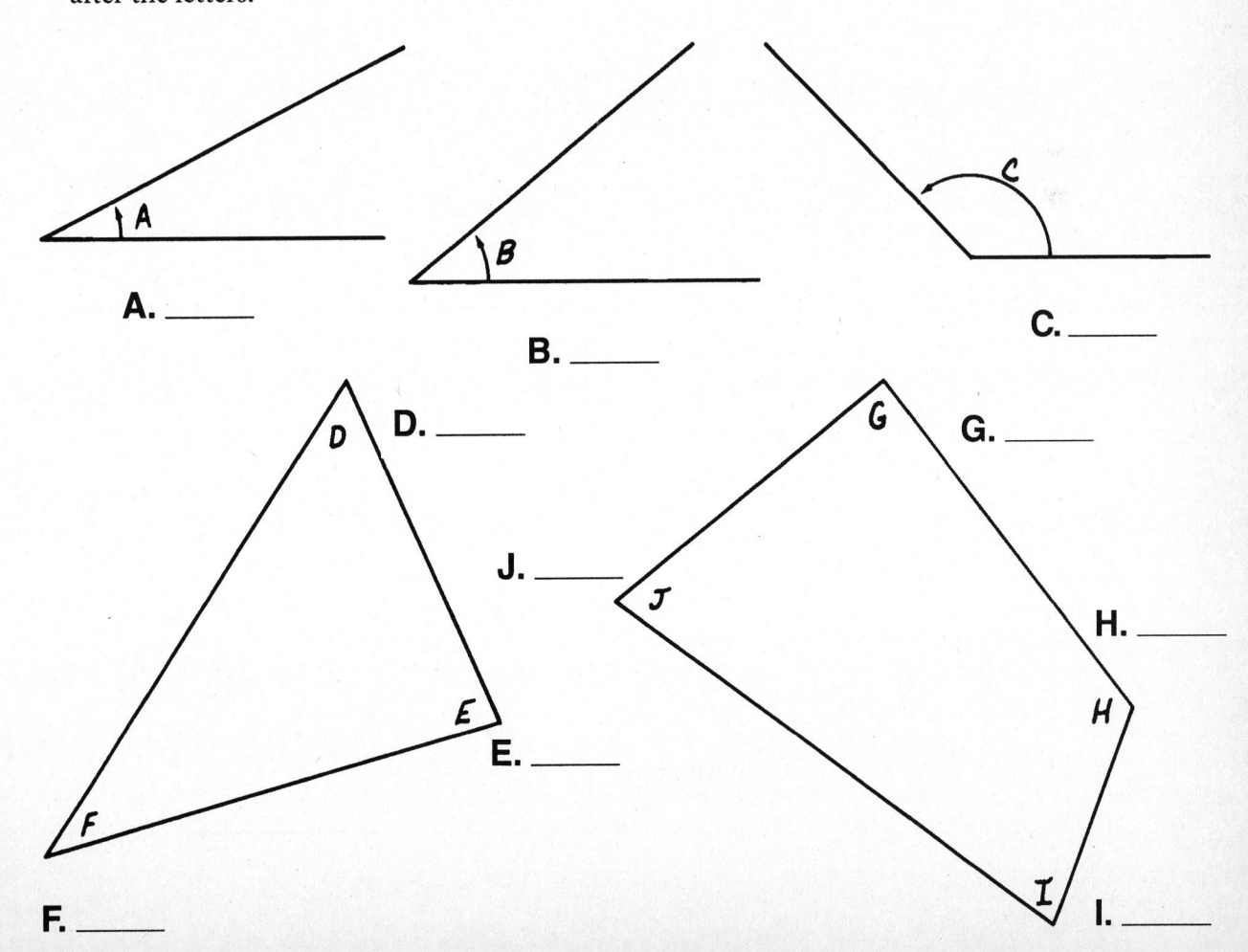

A. _____

B. _____

C. _____

D. _____

E. _____

F. _____

G. _____

H. _____

I. _____

J. _____

Copyright © Glencoe Division of Macmillan/McGraw-Hill

35

NAME _____ DATE _____ CLASS _____

LABORATORY SKILLS ASSESSMENT
Estimating Values Using SI

Lab and Safety Skills

The ability to estimate values such as length, mass, volume, and temperature is a useful skill. This skill can serve as a quick check of your answers to problems. For example, your bedroom is approximately 13 m² and your classroom is approximately 130 m². If you measured the length and width of your bedroom and calculated the area to be 130 m², an estimate of the answer would serve as a way of checking the placement of your decimal point.

For the questions in Part A, estimate the values using SI units for each example listed. Check your answers by either making the measurements yourself or using data your teacher provides. After answering the questions in Part A, estimate the values for the items listed in Part B and then check your answers. Compare your estimates in Parts A and B and answer the remaining questions.

A. Making Estimates

	Estimated values	Actual values

1. How long, tall, or far is it?
  a. a paper clip
  b. your teacher
  c. from Los Angeles to New York City
  d. from your home to your school

2. How much matter does it contain?
  a. one chocolate chip
  b. a soda cracker
  c. a newborn baby
  d. your science book

3. How much liquid will it hold?
  a. a school milk container
  b. a soft drink can
  c. a teaspoon
  d. a coffee cup

4. How hot or cold is it? (use °C)
  a. when water freezes
  b. your normal body temperature
  c. comfortable room temperature
  d. on a hot August day

5. For which type of estimating were your answers the most accurate? _____

6. For which type of estimating were your answers the least accurate? _____

7. Why do you think some types of estimating are easier for you than others? _____

NAME _____ DATE _____ CLASS _____

	Estimated values	Actual values

B. Making Estimates

1. How long, tall or far is it?
   a. a new pencil _____ _____
   b. a pine needle _____ _____
   c. a newborn baby _____ _____
   d. from Los Angeles to Dallas _____ _____

2. How much matter does it contain?
   a. a potato chip _____ _____
   b. a cat _____ _____
   c. a candy bar _____ _____
   d. a ten-speed bicycle _____ _____

3. How much liquid will it hold?
   a. a drinking glass _____ _____
   b. a straw _____ _____
   c. a baby bottle _____ _____
   d. a punch bowl _____ _____

4. How hot or cold is it? (Use °C)
   a. when water boils _____ _____
   b. inside your refrigerator _____ _____
   c. a cup of hot tea _____ _____
   d. during a snowstorm _____ _____

5. For which types of values did your estimating improve the most? _____
   _____
   _____
   _____

6. List three ways you can use estimating to help you in everyday activities. _____
   _____
   _____

7. What other activities could you do to help develop your estimating skills? _____
   _____
   _____
   _____

Copyright © Glencoe Division of Macmillan/McGraw-Hill

LABORATORY SKILLS ASSESSMENT
Scientific Problem Solving

Scientific Methods

All scientists have special interests. Scientists are interested in the world around them. This curiosity leads them to investigate things and events. Scientists use their senses to observe as they investigate. They use many methods in seeking answers to problems. Scientists use scientific problem solving. One scientific problem-solving technique has six steps:

1. State the problem.
2. Gather information about the problem.
3. Form a hypothesis.
4. Test the hypothesis.
5. Accept or reject the hypothesis.
6. Do something with the results.

Using Scientific Methods

Scientists observed that white mice that were fed seeds appeared to grow more than mice given leafy green and yellow vegetables. The scientists hypothesized that the protein in the seed was responsible for the growth. They designed an experiment to test this hypothesis. They divided 200 mice of the same age, size, health, and sex into two groups of 100 mice each. The mice were kept under identical conditions for fourteen days. One group was given a diet low in protein. The other group was given a normal protein diet. The mass of each mouse was recorded daily for fourteen days.

1. Which group of mice served as a control? the mice on a normal protein diet

2. What was the variable? the protein in the food

3. What effect of the protein diet was tested? the effect of different amounts of protein on growth in terms of mass

4. What other effects of a protein diet could have been tested? types or sources of different types of protein

5. Why were larger numbers of mice used in this experiment? the larger the number of individuals in a test, the more accurate the data and observations

6. If the results of the experiment did not show a marked change between the two groups, what should the scientists do next? rework the hypothesis and test another variable

7. What are the parts of an experiment? problem, procedure, observation, conclusion

LABORATORY SKILLS ASSESSMENT
Scientific Problem Solving

Tables and Graphs
A. Making Tables

As you study science, you will find that often information is presented in a table. You will organize data and observations in tables. You will be asked to use information given in tables. Knowing how to make and use tables is an important skill.

Tables have a title, rows, columns, and heads. The title is found at the top of the table. The title tells you what information is contained in the table. Columns are the sections that run up and down. At the top of each column is a head that tells you what information is in the column. Rows are sections that run from one side to another on the table.

Use the information in this paragraph to complete the table below. When you complete a table, you must record the information in the proper column and row. There are 5000 species in Phylum Porifera, 11 000 species in Phylum Coelenterata, 26 000 species in the three phyla of worms, 80 000 species in Phylum Mollusca, 826 000 species in Phylum Arthropoda, 47 000 species in Phylum Chordata, and 5000 species in Phylum Echinodermata.

Number of animal species	
Phylum	Number
Porifera	5000
Coelenterata	11 000
Worm Phyla	26 000
Mollusca	80 000
Arthropoda	826 000
Chordata	47 000
Echinodermata	5000

B. Making Graphs

A graph is a picture that shows data in a way that helps you understand the information. There are several different types of graphs. Line graphs show data plotted as points that are connected by a line. Line graphs often are used to show change. The graph in Figure 1 shows the increase in world population over time.

FIGURE 1

A bar graph always has two axes, a horizontal and a vertical, and these axes are labeled and divided. On the bar graph in Figure 2, the vertical axis is labeled and divided into ten thousands. The horizontal axis is labeled "Phylum" and divided by different animal phyla. Indicate the number of species in each phylum given by shading in bars on the graph. The column for Phylum Porifera has been done for you.

FIGURE 2

Pie graphs are circular graphs that show how each part is related to the total. Use the data from the table below to make a pie graph. Calculate the fractional amount of time for each of the sections. The total circle will represent 4.5 billion years. To find the fraction for each section, divide the time length of that section by 4 500 000 000. Then multiply the fraction by 360 degrees to determine the angle for the section. The calculation for the Cenozoic Era is given.

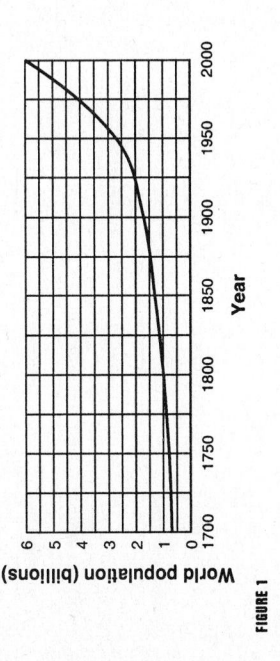

65 000 000 years ÷ by 4 500 000 000 years = 0.0144 × 360° = 5.18°

Round the final answer to the nearest whole degree. Use this procedure to complete the table. Then, use a protractor and the data from the table to construct a circular graph in Figure 3 on page 21. Label all parts of the graph.

Era	Number of years	Number of degrees
Cenozoic	65 000 000	5
Mesozoic	160 000 000	13
Paleozoic	345 000 000	27
Precambrian	3 930 000 000	314

Use a protractor and the data to construct a circular graph in Figure 3. Label all parts of the graph.

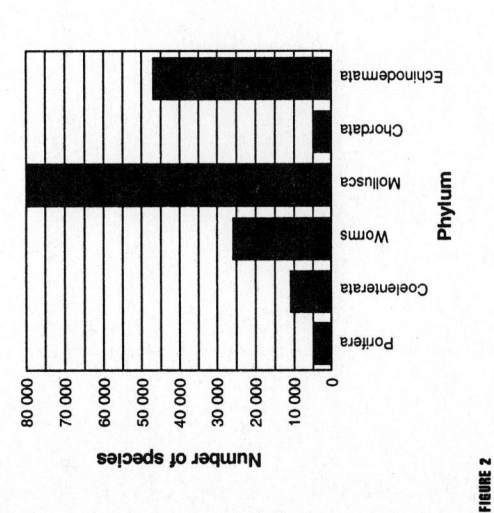

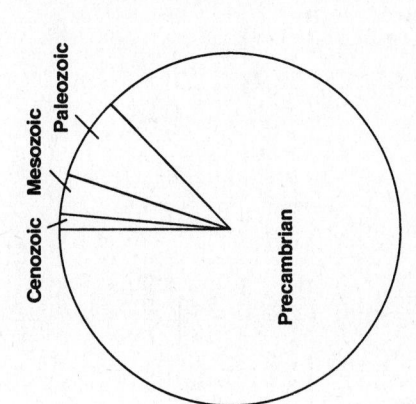

FIGURE 3

LABORATORY SKILLS ASSESSMENT

Lab and Safety Skills

Safety in the Science Classroom

A. Developing Safety Policies

Several statements are printed in Column I concerning students' activities and attitudes in the laboratory. Think about each statement and formulate a safety rule or procedure related to each statement. Write one or two clear, concise sentences in Column II that can serve as a safety policy for your science classroom. Use Appendix C in your textbook as a reference.

Column I

1. Peter says that his teacher is solely responsible for preventing laboratory accidents.

2. Keshia started the lab activity before reading it through completely.

3. Ricardo decided to do a lab activity that he read in a library book before the teacher came into the classroom.

4. Stephanie says that the lab aprons are unattractive and that the safety goggles mess up her hair. She refuses to wear them.

Column II

1. Everyone is responsible for safety. Pay attention to what you're doing. Be considerate of others. Don't fool around.

2. Lack of knowledge may cause an accident. Read through procedures carefully.

3. Perform activities only when a teacher is present and do only those activities authorized by a teacher.

4. Protect clothing and eyes by wearing laboratory aprons and goggles.

B. Using Safety Devices Correctly

Describe the location and purpose of having each of the following devices in your science laboratory. Write your answers in the spaces provided. (Locations will vary according to your laboratory design.)

1. fire blanket   When hair or clothing is on fire, apply blanket from head down to prevent fanning flames upward.

2. CO_2 fire extinguisher   Use on all fires except those involving people. Frostbite can result from skin contact with solid carbon dioxide.

3. goggles   wear when using any chemical

4. eyewash station or fountain   use for foreign material in the eye

5. safety hood or vent   Any procedures involving poisonous or irritating fumes should be done under a safety hood.

C. Identifying Improper Safety Practices

The pictures below show students performing laboratory activities incorrectly. Study each picture and write in the space provided all improper laboratory techniques that are illustrated. Be prepared to explain why it is important to follow each safety procedure.

1. Work in an uncluttered area; tie long hair back; wear safety goggles and apron; do not point open end of test tube at anyone; hold beaker while stirring.

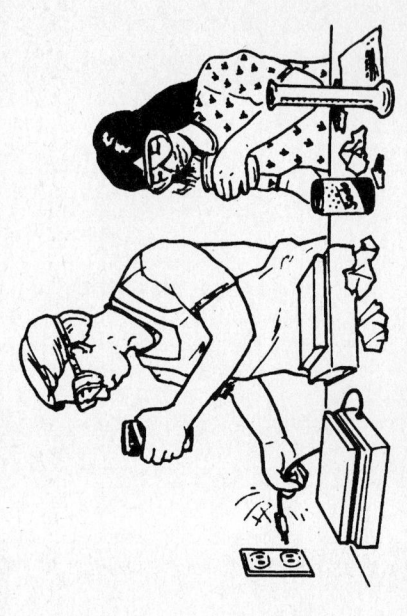

2. Do not eat or drink anything in the laboratory; use fanning motion to detect odors; do not pull electrical plug by the cord; work in an uncluttered area.

Explain what is wrong in each of the following diagrams and tell how to correct it.

3. Error ___stirring with a thermometer___

Correction ___Student should use a stirring rod.___

4. Error ___trying to read the liquid volume level from an angle___

Correction ___Student should view the meniscus of the liquid at eye level.___

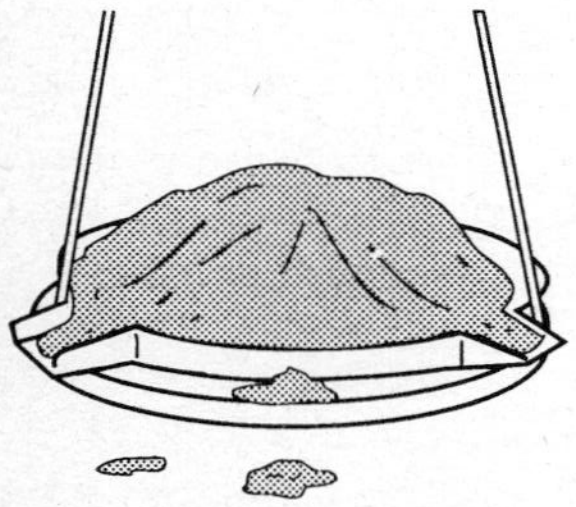

5. Error ___placing solid directly on the balance pan___

Correction ___Student should use waxed paper or a container.___

6. Error ___pushing straight down on tubing while holding it near the top___

Correction ___Student should use towels to protect hands, hold glass near stopper, and use twisting motion.___

7. Error ___carrying microscope incorrectly; base and stage could fall down; mirror might fall out___

Correction ___Student should use one hand to grasp the arm and use other hand under the base.___

8. Error ___handling a live animal with bare hands___

Correction ___Student should use gloves when handling a live animal to avoid being bitten.___

41

LABORATORY SKILLS ASSESSMENT
Using a Microscope

Lab and Safety Skills

Microscopes are optical instruments, much like a pair of glasses. The purpose of the microscope is to help you see things that are very small. The microscope can introduce you to organisms that you would otherwise not see. Because it is a delicate instrument, a microscope requires careful handling.

Show that you are aware of how to handle a microscope correctly by completing the following statements.

1. Always carry a microscope with one hand on the **arm** and the other hand on the **base**.
2. A microscope should be stored with the **low-power** objective in place.
3. Always bring a specimen into focus using the **low-power** objective.
4. Never use the **coarse** adjustment to focus the high-power objective.
5. Do not allow direct sunlight to fall on the **mirror**.
6. Use only the **fine** adjustment when focusing with the high-power objective.
7. Lenses should be cleaned with **lens paper**.

Identify the parts of the microscope on the lines to the left.

a. **eyepiece**
b. **body tube**
c. **revolving nosepiece**
d. **low-power objective lens**
e. **high-power objective lens**
f. **stage**
g. **stage clips**
h. **base**
i. **mirror**
j. **diaphragm**
k. **arm**
l. **fine adjustment knob**
m. **coarse adjustment knob**

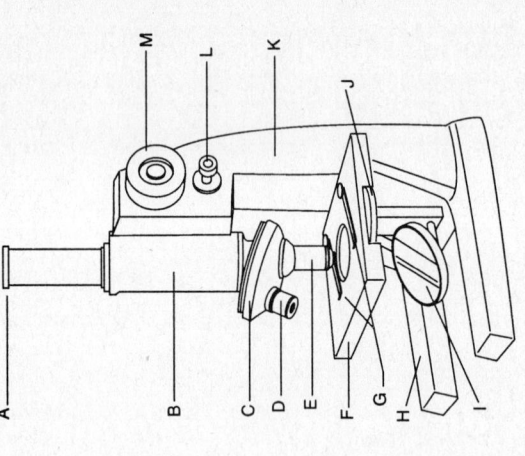

LABORATORY SKILLS ASSESSMENT
Measuring with a Microscope

Lab and Safety Skills

Scientists use microscopes to study things that are too small to be seen with the unaided eye. The size of microscopic organisms can be estimated using the method given here. The approximate size of the field of view seen under low power can be determined by actual measurement. A transparent millimeter ruler is placed across the field of view as shown in Figure 1.

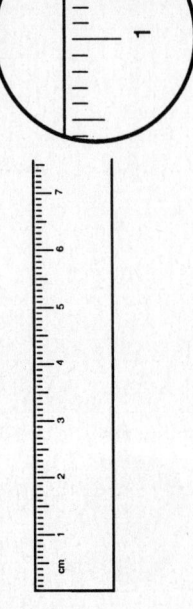

FIGURE 1

1. How many millimeters wide is the field of view shown? **about 12 mm**

Objects examined under the microscope are quite small. Therefore, it is often convenient to use units of length smaller than the millimeter for microscope measurement. One unit often used is the micron (μ). There are 1000 microns in one millimeter.

2. How many microns wide is the low-power field shown in Figure 1? **12 000 μ (microns)**

To measure the diameter of the high-power field of view, follow these steps. First divide the magnification number of the high-power objective by the magnification number of the low-power objective. Then divide the diameter of the low-power field of view by this number. For example, assume that a low-power objective is 10×, and the higher power objective is 40×. On this microscope, the diameter of the low-power field of view is 2000 microns.

3. What is the width of the high-power field of view? **300 microns**

4. What is the width of the organisms shown below? These organisms were viewed under high power.

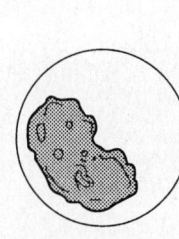

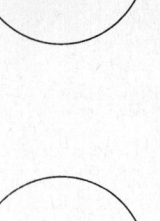

200 microns   125 microns   150 microns

FIGURE 2

LABORATORY SKILLS ASSESSMENT
Using Laboratory Measuring Devices

Lab and Safety Skills

Balance: Determining Mass
1. What mass is shown on each of these balances?

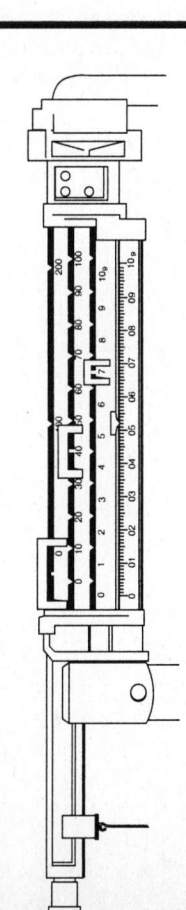

a. The mass of the object would be read as **47.52 g**.

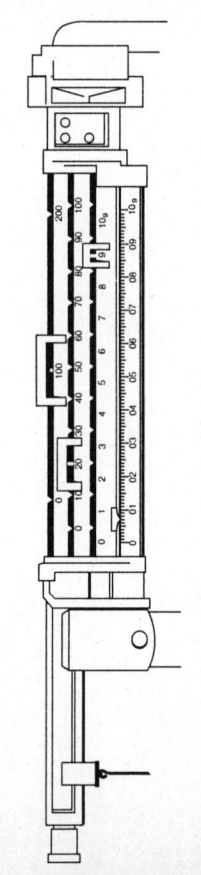

b. The mass of the object would be read as **129.07 g**.

Metric Ruler: Determining Length
2. What lengths are indicated on this ruler?

a. 12.50 cm
b. 11.30 cm
c. 13.25 cm
d. 13.95 cm
e. 10.30 cm
f. 12.00 cm
g. 9.70 cm
h. 12.96 cm
i. 14.34 cm
j. 11.71 cm

Graduated Cylinder: Measuring Liquid Volume
3. What volume is indicated on each of these graduated cylinders?

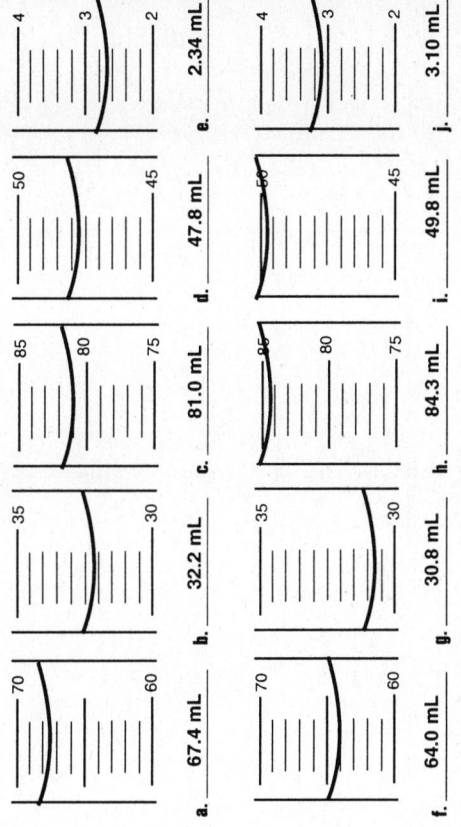

a. 67.4 mL
b. 32.2 mL
c. 81.0 mL
d. 47.8 mL
e. 2.34 mL
f. 64.0 mL
g. 30.8 mL
h. 84.3 mL
i. 49.8 mL
j. 3.10 mL

Thermometer: Measuring Temperature
4. What temperature is indicated on each of these thermometers?

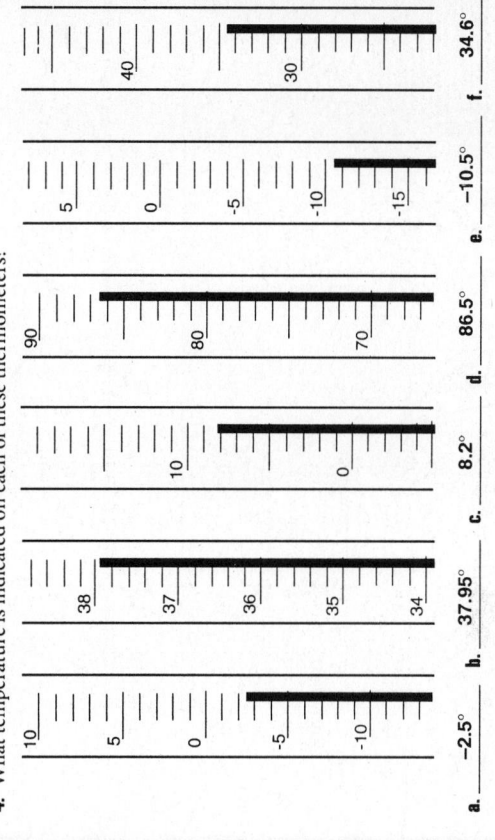

a. −2.5°
b. 37.95°
c. 8.2°
d. 86.5°
e. −10.5°
f. 34.6°

LABORATORY SKILLS ASSESSMENT
Using a Metric Ruler and Protractor

Lab and Safety Skills

A. Calculating Surface Area and Volume Using Metric Measurements

Your teacher will give you several objects including those items listed in Table 1. Using your metric ruler, make the required measurements and complete the table. Use your data to calculate surface area or volume for each item.

TABLE 1

Object	Length	Width	Height	Surface area/volume
Index card	12.8 cm	7.6 cm		$A = l \times w$ $A = 97$ cm²
Microscope slide	7.5 cm	2.5 cm	0.080 cm	$V = l \times w \times h$ $V = 1.5$ cm³
Petri dish	diameter = 6.0 cm radius = 3.0 cm		1.5 cm	$V = \pi r^2 h$ $V = 42$ cm³
Chalk	diameter = 1.2 cm radius = 0.60 cm		8.5 cm	$V = \pi r^2 h$ $V = 9.6$ cm³
Desk top	Answers will vary.			

B. Metric Scale Conversions

Using your metric ruler and the map on page 35, measure the distances between the points indicated in Table 2. Record your answers in Table 2. Complete the table using the scale 10 cm = 1 km.

TABLE 2

How far is it from the _____?	Metric measurement mm	Metric measurement cm	Actual distance km
Softball field to the lake	130	13.0	1.30
Meadow to the picnic area	45	4.5	0.45
Hickory grove to the lake	57	5.7	0.57
Pine forest to the picnic area	71	7.1	0.71
Softball field to the hickory grove	93	9.3	0.93
Softball field to the lake through the meadow	66 + 73 = 139	13.9	1.39

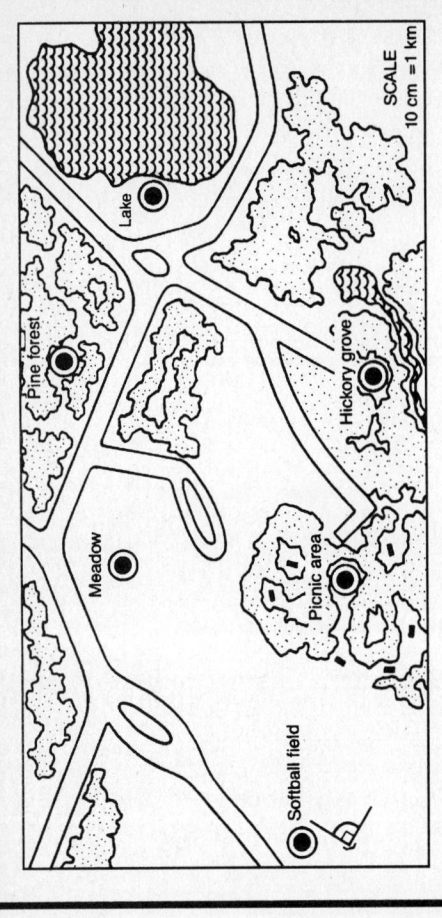

C. Measuring Angles

1. Using your protractor, measure the size of the angles below. Record your answers on the lines after the letters.

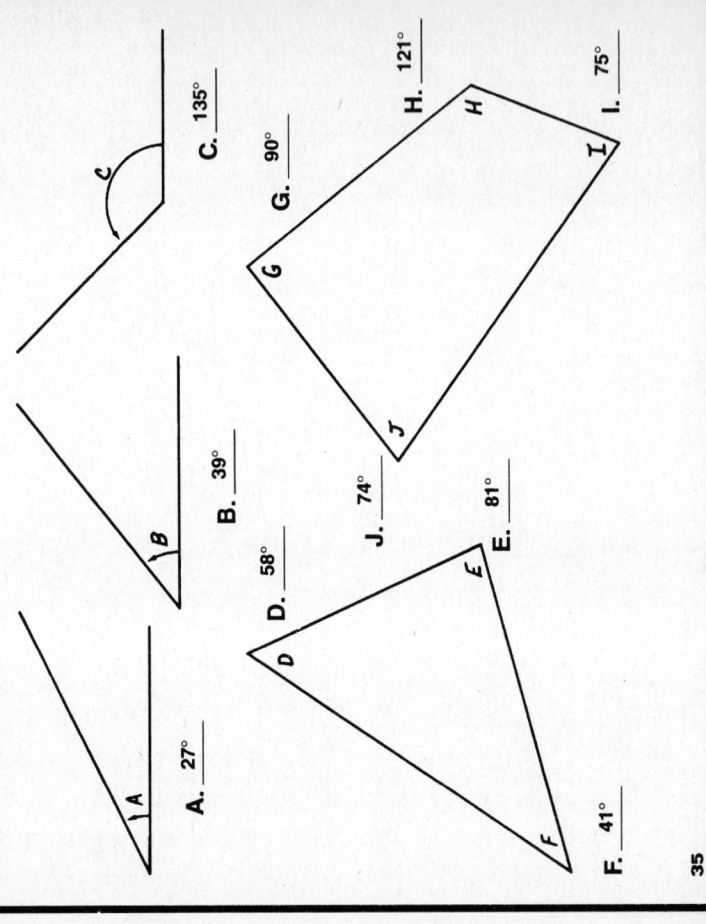

A. 27°
B. 39°
C. 135°
D. 58°
E. 81°
F. 41°
G. 90°
H. 121°
I. 75°
J. 74°

LABORATORY SKILLS ASSESSMENT
Estimating Values Using SI

Lab and Safety Skills

The ability to estimate values such as length, mass, volume, and temperature is a useful skill. This skill can serve as a quick check of your answers to problems. For example, your bedroom is approximately 13 m² and your classroom is approximately 130 m². If you measured the length and width of your bedroom and calculated the area to be 130 m², an estimate of the answer would serve as a way of checking the placement of your decimal point.

For the questions in Part A, estimate the values using SI units for each example listed. Check your answers by either making the measurements yourself or using data your teacher provides. After answering the questions in Part A, estimate the values for the items listed in Part B and then check your answers. Compare your estimates in Parts A and B and answer the remaining questions.

A. Making Estimates

	Estimated values	Sample data Actual values
1. How long, tall, or far is it?	Answers will vary	
a. a paper clip		3.2 cm
b. your teacher		1.8–2.1 m
c. from Los Angeles to New York City		3600 km
d. from your home to your school		Answers will vary
2. How much matter does it contain?		
a. one chocolate chip		0.5 g
b. a soda cracker		10 g
c. a newborn baby		2–5 kg
d. your science book		0.5 g
3. How much liquid will it hold?		
a. a school milk container		0.25 L
b. a soft drink can		0.33 L
c. a teaspoon		5 mL
d. a coffee cup		0.20 L
4. How hot or cold is it? (use °C)		
a. when water freezes		0°C
b. your normal body temperature		37°C
c. comfortable room temperature		25°C
d. on a hot August day		30°–35°C

5. For which type of estimating were your answers the most accurate? __Answers will vary.__

6. For which type of estimating were your answers the least accurate? __Answers will vary.__

7. Why do you think some types of estimating are easier for you than others? __Answers will vary.__

B. Making Estimates

	Estimated values	Sample data Actual values
1. How long, tall or far is it?	Answers will vary	
a. a new pencil		19 cm
b. a pine needle		6 cm
c. a newborn baby		50–55 cm
d. from Los Angeles to Dallas		1800 km
2. How much matter does it contain?		
a. a potato chip		1.5–2.0 g
b. a cat		2 kg
c. a candy bar		40–55 g
d. a ten-speed bicycle		15 kg
3. How much liquid will it hold?		
a. a drinking glass		0.33 L
b. a straw		2 mL
c. a baby bottle		0.25 L
d. a punch bowl		4 L
4. How hot or cold is it? (Use °C)		
a. when water boils		100°C
b. inside your refrigerator		5°C
c. a cup of hot tea		55°C
d. during a snowstorm		0°C or lower

5. For which types of values did your estimating improve the most? __Answers will vary.__

6. List three ways you can use estimating to help you in everyday activities. __Answers will vary.__

7. What other activities could you do to help develop your estimating skills? __Answers will vary.__